MORPHOLOGICAL COMPLEXITY

Inflectional morphology plays a paradoxical role in language. On the one hand, it tells us useful things, for example that a noun is plural or a verb is in the past tense. On the other hand, many languages get along perfectly well without it, so the baroquely ornamented forms we sometimes find come across as a gratuitous over-elaboration. This is especially apparent where the morphological structures operate at cross-purposes to the general systems of meaning and function that govern a language, yielding inflection classes and arbitrarily configured paradigms. This is what we call *morphological complexity*. Manipulating the forms of words requires learning a whole new system of structures and relationships. This book confronts the typological challenge of characterizing the wildly diverse sorts of morphological complexity we find in the languages of the world, offering both a unified descriptive framework and quantitative measures that can be applied to such heterogeneous systems.

MATTHEW BAERMAN is Senior Research Fellow in the Surrey Morphology Group at the University of Surrey, whose work concentrates on the description, typology, and diachrony of morphology, in particular complex inflectional systems. He is the editor of the recent *Oxford Handbook of Inflection* (2015).

DUNSTAN BROWN is Professor and Head of the Department of Language and Linguistic Science, University of York, and a visiting professor in the Surrey Morphology Group. Recent publications include: *Network Morphology* (with Andrew Hippisley, 2012); and as co-editor, *Canonical Morphology and Syntax* (2012), *Understanding and Measuring Morphological Complexity* (2015) and *Archi: Complexities of Agreement in Cross-Theoretical Perspective* (2016).

GREVILLE G. CORBETT is Distinguished Professor of Linguistics, University of Surrey, where he leads the Surrey Morphology Group. He works on the typology of features, as in the previously published *Gender* (1991), *Number* (2000), *Agreement* (2006), and *Features* (2012), all with Cambridge University Press. He is a fellow of the British Academy and of the Academy of Social Sciences, a member of the Academia Europaea and an honorary member of the Linguistic Society of America.

In this series

116. GILLIAN CATRIONA RAMCHAND: *Verb Meaning and the Lexicon: A First Phase Syntax*
117. PIETER MUYSKEN: *Functional Categories*
118. JUAN URIAGEREKA: *Syntactic Anchors: On Semantic Structuring*
119. D. ROBERT LADD: *Intonational Phonology* (second edition)
120. LEONARD H. BABBY: *The Syntax of Argument Structure*
121. B. ELAN DRESHER: *The Contrastive Hierarchy in Phonology*
122. DAVID ADGER, DANIEL HARBOUR, and LAUREL J. WATKINS: *Mirrors and Microparameters: Phrase Structure beyond Free Word Order*
123. NIINA NING ZHANG: *Coordination in Syntax*
124. NEIL SMITH: *Acquiring Phonology*
125. NINA TOPINTZI: *Onsets: Suprasegmental and Prosodic Behaviour*
126. CEDRIC BOECKX, NORBERT HORNSTEIN and JAIRO NUNES: *Control as Movement*
127. MICHAEL ISRAEL: *The Grammar of Polarity: Pragmatics, Sensitivity, and the Logic of Scales*
128. M. RITA MANZINI and LEONARDO M. SAVOIA: *Grammatical Categories: Variation in Romance Languages*
129. BARBARA CITKO: *Symmetry in Syntax: Merge, Move and Labels*
130. RACHEL WALKER: *Vowel Patterns in Language*
131. MARY DALRYMPLE and IRINA NIKOLAEVA: *Objects and Information Structure*
132. JERROLD M. SADOCK: *The Modular Architecture of Grammar*
133. DUNSTAN BROWN and ANDREW HIPPISLEY: *Network Morphology: A Defaults-Based Theory of Word Structure*
134. BETTELOU LOS, CORRIEN BLOM, GEERT BOOIJ, MARION ELENBAAS and ANS VAN KEMENADE: *Morphosyntactic Change: A Comparative Study of Particles and Prefixes*
135. STEPHEN CRAIN: *The Emergence of Meaning*
136. HUBERT HAIDER: *Symmetry Breaking in Syntax*
137. JOSÉ A. CAMACHO: *Null Subjects*
138. GREGORY STUMP and RAPHAEL A. FINKEL: *Morphological Typology: From Word to Paradigm*
139. BRUCE TESAR: *Output-Driven Phonology: Theory and Learning*
140. ASIER ALCÁZAR and MARIO SALTARELLI: *The Syntax of Imperatives*
141. MISHA BECKER: *The Acquisition of Syntactic Structure: Animacy and Thematic Alignment*
142. MARTINA WILTSCHKO: *The Universal Structure of Categories: Towards a Formal Typology*
143. FAHAD RASHED AL-MUTAIRI: *The Minimalist Program: The Nature and Plausibility of Chomsky's Biolinguistics*
144. CEDRIC BOECKX: *Elementary Syntactic Structures: Prospects of a Feature-Free Syntax*
145. PHOEVOS PANAGIOTIDIS: *Categorial Features: A Generative Theory of Word Class Categories*
146. MARK BAKER: *Case: Its Principles and Its Parameters*
147. WILLIAM BENNETT: *The Phonology of Consonants: Dissimilation, Harmony, and Correspondence*
148. ANDREA SIMS: *Inflectional Defectiveness*
149. GREGORY STUMP: *Inflectional Paradigms: Content and Form at the Syntax-Morphology Interface*
150. ROCHELLE LIEBER: *English Nouns: The Ecology of Nominalization*
151. JOHN BOWERS: *Deriving Syntactic Relations:*
152. ANA TERESA PEREZ-LEROUX: *The Acquisition of Complex Noun Phrases*
153. MATTHEW BAERMAN, DUNSTAN BROWN, and GREVILLE G. CORBETT: *Morphological Complexity*

Earlier issues not listed are also available

CAMBRIDGE STUDIES IN LINGUISTICS

Morphological Complexity

MORPHOLOGICAL COMPLEXITY

MATTHEW BAERMAN
University of Surrey

DUNSTAN BROWN
University of York

GREVILLE G. CORBETT
University of Surrey

CAMBRIDGE
UNIVERSITY PRESS

CAMBRIDGE
UNIVERSITY PRESS

University Printing House, Cambridge CB2 8BS, United Kingdom

Cambridge University Press is part of the University of Cambridge.

It furthers the University's mission by disseminating knowledge in the pursuit of education, learning, and research at the highest international levels of excellence.

www.cambridge.org
Information on this title: www.cambridge.org/9781107120648
DOI: 10.1017/9781316343074

First published 2017

A catalogue record for this publication is available from the British Library.

ISBN 978-1-107-12064-8 Hardback

Contents

Figures

Tables

Acknowledgements

For helpful comments at different stages of the development of this book we are grateful to: Olivier Bonami, Oliver Bond, Gilles Boyé, Wayles Browne, Patrica Cabredo Hofherr, Marina Chumakina, Scott Collier, Roger Evans, Sebastian Fedden, Tim Feist, Jürg Fleischer, Alexander Krasovitsky, Michele Loporcaro, Steve Marlett, Irina Monich, Enrique Palancar, Ingo Plag, Ljubomir Popović, Bert Remijsen, David Robinson, Serge Sagna, Benoît Sagot, Andrew Spencer, and Anna Thornton. Versons of parts of this research have been presented by each of us at different conferences and seminars: we greatly appreciate all the suggestions offered there. The Nuer examples in Chapter 4 were kindly provided by John Gai, Yak Wichok, and Lam Muang. We would also like to thank Penny Everson for assistance in the preparation of the manuscript.

This work has been funded by various research councils, whose support is gratefully acknowledged: the European Research Council (ERC-2008-AdG-230268: MORPHOLOGY "Morphological Complexity"), Arts and Humanities Research Council (AH/I027193/1: "From Competing Theories to Fieldwork: The Challenge of an Extreme Agreement System"), Economic and Social Research Council & Arts and Humanities Research Council (ES/I029621/1: "Endangered Complexity: Inflectional Classes in Oto-Manguean Languages"), and the Arts and Humanities Research Council (AH/L011824/1: "Morphological Complexity in Nuer" and AH/N006887/1: "Lexical splits: a novel perspective on the structure of words").

Abbreviations

1	first person
2	second person
3	third person
ABL	ablative
ABS	absolutive
ACC	accusative
ADV	adverbial
ALL	allative
AOR	aorist
COM	comitative
COND	conditional
CONT	continuous
COREF	coreferential
CPL	completive
CSN	comparison
DAT	dative
DEP	dependent
DO	direct object
DPP	dynamic principal parts ratio
DU	dual
DUB	dubitative
ERG	ergative
EXCL	exclusive
F	feminine
FUT	future
GEN	genitive
GER	gerund
HAB	habitual
HORT	hortative
IMP	imperative

IMPRF	imperfect
INCL	inclusive
INCPL	incompletive
IND	indicative
INF	infinitive
INS	instrumental
IPFV	imperfective
IRR	irrealis
JUSS	jussive
La	Latin
LAT	lative
LOC	locative
M	masculine
N	neuter
NEG	negative
NOM	nominative
NPST	non-past
PART	partitive
PFV	perfective
PL	plural
POT	potential
PPRF	pluperfect
PRF	perfect
PROG	progressive
PROSP	prospective
PRS	present
PST	past
REAL	realis
RECIP	recipient
RPRO	resumptive pronoun
SBJ	subject
SBJV	subjunctive
SG	singular
SIMUL	simultaneous
Skr	Sanskrit
TAM	tense-aspect-mood
TR	transitive
VOC	vocative

1 *Introduction*

Inflectional morphology equips a language with the means to combine lexical and grammatical information. For example, the Aymara word *anunakataki* and the Polish word *psom* both mean 'for dogs', and in each language the notions of being a dog, plurality, and benefaction are expressed by a single word form. By contrast, in a language like Tok Pisin, each of these three notions is conveyed by a separate word: *long* (derived from English *belong*, expressing benefaction), *ol* (derived from English *all*, expressing plurality), and *dok* (derived from *dog*, the lexical information); these combine into the phrase *long ol dok*. As a medium of expression, the morphological forms of Aymara and Polish are doing the same job as the syntactic construction of Tok Pisin, and the only difference is how the information is packaged: as separate words (Tok Pisin), or as aspects of a single word (Aymara and Polish). But the inclusion of this information in the word form opens up a new dimension – the paradigm – absent in purely syntactic arrangements. And with this new dimension come new relationships and a host of complexities characteristic of morphological systems.

By way of illustration, consider the noun paradigms in Table 1.1, from Aymara, an Aymaran language spoken in Bolivia and neighbouring countries, and in Table 1.2, from the Slavonic language Polish. The Aymara paradigms differ only minimally from what we might construct, item-by-item, in syntax. We can identify a noun stem and a set of elements that follow, and then concatenate the two. Morphophonological quirks show us that this is not *just* syntactic concatenation: the accusative and comparative both require deletion of the stem-final vowel found in most of the rest of the paradigm. But aside from this essentially implementational detail, the organization of these paradigms is much the same as it would be if they were bare nouns paired with various different adpositions: in order to construct an ablative case form meaning 'from the house', we take the stem labelled 'house' (*uta*) and the suffix labelled ABLATIVE (*-ta*) in the right column, and put the two together.

Table 1.1 *Aymara noun paradigms*

	'house'	'money'	'dog'	suffixes
NOMINATIVE	uta	qullqi	anu	Ø
ACCUSATIVE	ut	qullq	an	Ø
GENITIVE	uta-na	qullqi-na	anu-na	na
ABLATIVE	uta-ta	qullqi-ta	anu-ta	ta
ALLATIVE	uta-ru	qullqi-ru	anu-ru	ru
INSTRUMENTAL	uta-mpi	qullqi-mpi	anu-mpi	mpi
BENEFACTIVE	uta-taki	qullqi-taki	anu-taki	taki
COMPARATIVE	ut-hama	qull-hama	an-hama	hama
INTERACTIVE	uta-pura	qullqi-pura	anu-pura	pura
LIMITATIVE	uta-kama	qullqi-kama	anu-kama	kama
PURPOSIVE	uta-lajku	qullqi-lajku	anu-lajku	lajku

Source: Coler 2015.

Table 1.2 *Polish noun paradigms (singular forms)*

	'fish'	'nose'	'cart'	suffixes		
NOMINATIVE	ryb-a	nos	wóz	a	Ø	Ø
ACCUSATIVE	ryb-ę	nos	wóz	ę	Ø	Ø
GENITIVE	ryb-y	nos-a	woz-u	y	a	u
DATIVE	rybi-e	nos-u	woz-u	e	u	u
LOCATIVE	rybi-e	nosi-e	wozi-e	e	e	e
INSTRUMENTAL	ryb-ą	nos-em	woz-em	ą	em	em

Now consider the Polish paradigms in Table 1.2. Here too we have three different nouns. But we cannot assemble these forms with the same ease as in Aymara. First, instead of there being a single set of suffixes that can be matched to any noun, there are different suffixes that occur with different nouns: for example, 'fish' takes a genitive singular suffix *-y*, whereas 'nose' takes *-a*. Simply labelling *-y* as genitive singular and *-a* as genitive singular will not tell us when to use which. Second, the way the grammatical functions themselves are packaged differs between the nouns: for example, the suffix *-u* is used for dative singular with 'nose', but for both genitive and dative with 'cart'. That means no single grammatical label will tell us when exactly to use *-u*.

The Polish word forms are so shot through with lexical idiosyncracies that it takes more work to say what is going on than with Aymara. Schematically, we can represent the contrasting descriptive tasks as in Table 1.3. Following the same schema as in Tables 1.1 and 1.2, we list the morphosyntactic values (like NOMINATIVE or DATIVE) as rows, and the lexemes as columns.

Table 1.3 *Simplicity vs. complexity*

a. Aymara-type system

	lexeme 1,2,3
value a	
value b	
value c	

b. Polish-type system

<table>
<tr><td></td><td>lexeme 1</td><td>lexeme 2</td><td>lexeme 3</td></tr>
<tr><td>value a</td><td></td><td></td><td rowspan="2">↕</td></tr>
<tr><td>value b</td><td rowspan="2">↕</td><td></td></tr>
<tr><td>value c</td><td></td><td></td></tr>
</table>

For Aymara there is just one column, because all the lexemes behave in the same way, and we need only to fill in the blanks by means of rules. For Polish, we need to expand this model along the horizontal dimension, creating three columns to accommodate three patterns of inflection for three different classes of lexeme. And further, in the vertical dimension, we need to acknowledge that the dimensions of the cells vary across items, if we are to be true to the evidence provided by the forms themselves. This is what we mean by morphological complexity: on the assumption that the two systems in Table 1.3 are doing the same job, the Polish-type system is more complex, because we need more elements and more steps to describe it. We should, however, caution against overinterpreting our use of the term *complexity*. We take no stand here on its cognitive effects – that is, is it really complex for the language user? – or on the possibility that it may have some application above and beyond the grammatical functions that define the paradigm – for example, as an aid to memory. This approach to complexity is thus local and not global, in the terms discussed by Miestamo (2006, 2008), in that it is defined in relation to a particular subsystem. This is not because the broader systemic implications are not interesting or not relevant, but because the main aim of this volume is typological, to explore what kinds of configurations are actually found in languages. To that end, our terms of analysis are those that are generally used for inflectional systems, even for otherwise poorly described languages; this typically means an inventory of forms and the contexts they are found in.

Complexity in this sense is not just an interesting property of a subset of morphological systems. It is, more generally, what sets morphology apart from other linguistic components. It is restricted to paradigmatic structures: the whole idea that morphosyntactic values can be discussed separately from their means and patterns of exponence presupposes a paradigmatic arrangement of facts. And paradigms are characteristically morphological, and in particular, inflectional.[1] In effect, morphological complexity is the defining property of morphology as an autonomous linguistic component. Our aim here is to confront the full richness of morphological complexity, accepting that these patterns take on a morphological life of their own, worth exploring for their elegance and interest, and especially since they are arguably the most purely linguistic part of language.

This puts us in a somewhat awkward position, which we freely admit. On the one hand, we have identified our topic in terms of inflectional patterns, but at the same time, we have defined these patterns negatively, as what is left over after morphosyntax has been subtracted. Equally, most observers would ascribe inflectional patterns that coincide semantic or phonological regularities to those very modules. Morphological structure is the explanation of last resort, invoked when all other means of analysis have been exhausted. As a heuristic, this is only right and proper. Syntax, semantics, phonology – these are aspects of every language, and no description, whatever framework it is embedded in, can get by without making reference to them. But inflection, let alone what we have described as morphological complexity, is not found in every language, and so it understandably assumes a subordinate role in our thinking. This sets the bar rather high for identifying inflectional structure as a linguistic phenomenon in its own right. First, it must meet the paradigmatic criteria illustrated in Table 1.3. But even then, so one argument goes, the surface patterns could be accidental, and not in any way officially sanctioned, in the same way that homophony and synonymy could be seen as quirks of the lexicon. Thus, a second criterion is that the patterns give some evidence of systematicity, for example, through diachronic persistence. As forcefully argued by Maiden (1992, 2005), morphological patterns whose origins lie in a chance association of disparate elements can nevertheless be propagated across generations and serve as a driver of morphological change, drawing new items into their sphere of influence. Maiden's examples are from Romance, and we supplement these with comparable examples from Finnic languages in Section 2.2. In the light of the sometimes contentious status of morphology within language, such case studies are crucial to demonstrating that somewhere, somehow, purely morphological relationships call the shots, at the expense, seemingly, of all considerations of meaning or function. However, it is not often that we can find such convincing cases, where the diachronic record is solid and where we

[1] We note that the notion of derivational paradigms has some currency at the moment. We make no claims here about the distinction between inflection and derivation – paradigmaticity is just one property among many that some use to make a distinction. Anything paradigmatic falls under our purview.

can distinguish productive patterns from mere inertia (if there is even a difference). Does that make autonomous morphological structure a real but rare phenomenon, like three degrees of vowel length or paucal number? Possibly, provided one accepts the primacy of non-morphological aspects of language. But we would like to suggest here that that is a hasty conclusion. As we hope to show in the ensuing chapters, languages abound in the sorts of paradigmatic deviations illustrated in Table 1.3. They are easily overlooked if one's focus is a morphosyntactic description, but these peculiarities are nonetheless crucial to understanding or producing an utterance in any language that contains them.

We are therefore taking a novel and somewhat eclectic approach to the topic. We are not trying to advance a particular theoretical claim, and so have aimed to maintain a presentation free from polemic. Our definition of morphological complexity is fluid in that it is dependent on what one assumes is already provided by other aspects of the linguistic system. But these other aspects – for example, syntax, semantics, phonology – are largely intangibles, as indeed is morphological structure itself. Our focus is therefore on what is most solidly observable – on forms and their distribution. The goal is to offer a typology of morphological complexity, gathering its various manifestations under one rubric. Armed with this, readers are, of course, free to decide where this fits in their own conceptions of grammar, we hope with an enhanced awareness of the formal and typolological richness of inflectional systems.

In concrete terms, this emerges as a study of inflection classes. This means not just variation in the shape of inflectional marking across different sets of lexemes, as traditionally understood, but also variation in the paradigmatic distribution of inflectional marking, as seen in the Polish example in Table 1.2, where the morphosyntactic value of the suffixes *-a*, *-e* and *-u* varies from paradigm to paradigm. One issue we have chosen largely to skirt around here is the conflation of values within the paradigm, otherwise known as syncretism. This is, partly because we have explored it before (Baerman, Brown, & Corbett 2005), and partly because it would be a distraction from our main goal, given the wealth of competing analyses of feature structure whose plausibility may be dependent on theory-specific assumptions. Distributional variation in the feature values of (what may be supposed to be) identical morphological formatives is a phenomenon which *may* overlap with that of syncretism (e.g., Polish *-u* is dative with 'nose' but genitive/dative with 'cart'), but it need not (e.g., *-a* is nominative with 'fish' and genitive with 'nose').

The volume has the following structure. Chapter 2 surveys the external typology of inflection classes. Although they are most clearly and uncontroversially manifested through affix allomorphy (as in the Polish examples just illustrated), there is every reason to extend the notion to other types of inflectional exponence, as well as to more abstract properties of inflectional structure, such as affix position, variations in the paradigmatic distribution of affixes, or indeed the very presence versus absence of inflectional marking. Chapter 3 looks from the inside out, as it were, in terms of the morphosyntactic features that are being marked. Although in theory any feature is prone to have its expression split into different inflection classes, there are hints that the way this is manifested can be affected by the semantics of the feature itself. Chapter 4 shifts to the central question of inflection class assignment. Canonically speaking, inflection classes are completely arbitrary, unconnected with any other grammatical component. But often matters are not so clear, with morphological behaviour predictable to some degree from external factors, such as phonology or semantics, showing there is a substantial grey zone in between what we might call motivated and arbitrary morphology. In Chapter 5 we focus on conditions on paradigms; these are generalizations which cross-cut the lexical generalizations which are inflection classes. These conditions, ranging from the very general to the quite specific, have an interesting and surprisingly complete typology. This typology is based on the antecedent of the condition (semantic, syntactic, morphological, or phonological) and the type of paradigm in the consequent of the condition. Chapter 6 steps back and looks at the abstract properties of paradigm structure that are revealed through inflection classes, in terms of predictability both across cells of the paradigm, and across different components of the inflectional system. We identify three basic types of systems. In Chapter 7 we show how these basic types can be associated with three different ways of viewing morphological complexity. The first of these is organization, where the system is easy to specify in the grammar. The second of these is the opposite of organization – emergent complexity, where the major burden is associated with lexical stipulation and there is virtually no role for the morphological grammar. The third is central-system complexity, the trade-off between the two other types, where morphological systems that are high in central system complexity represent a balance between stipulation in the lexicon and the grammatical-rule system. These three types are then associated with three off-the-shelf measures available from Stump & Finkel (2013). In real world systems, the situation is much more intricate, of course, and we provide a case study of Tlatepuzco Chinantec tone classes, illustrating the interplay of lexical and grammatical knowledge, as represented in the system of default-inheritance classes. Chapter 8 sums our overview of morphological complexity in general, and inflectional classes in particular.

A thread running throughout this volume is the cross-linguistic pervasiveness of morphological complexity as manifested through inflection class distinctions. Although the most familiar examples in the literature come from Indo-European languages, few language families that have inflectional morphology are entirely free from such quirks (and those that seem to be may well just be victims of our own ignorance). We have drawn our examples here from a wide range of languages spanning all the inhabited continents, representing typologically very different inflectional systems. That said, it must also be admitted that some language families are especially prone to morphological complexity, the Oto-Manguean family being a particular stand out, and we have tempered genetic balance with the sort of typological diversity that can be gained only from dipping repeatedly into the same well.

The types of linguistic structures that we are concerned with here are often overlooked in more general discussions of language for partly understandable reasons. They are not necessary for transmitting information, they are not necessary for the smooth running of the sound system. Indeed, they are simply not necessary (many languages have limited morphology). And yet they are there, complex, difficult, and persistent, as we shall see.

2 *External Typology of Inflection Classes*

The mention of *inflection classes* may conjure up very different pictures, depending on which languages one is familiar with. And in practice, the sorts of systems commonly characterized as inflection classes differ widely. This means that any serious engagement with the issue requires us to make clear in every instance exactly what it is we are talking about. We start by looking in section 2.1 at the ways inflection classes are manifested through affixation, turning then to other means of morphological exponence: stems (section 2.2), suprasegmental features (section 2.3), inflectedness versus uninflectedness (section 2.4), considering also the possibility that words themselves could be understood as exponents of inflection and hence bearers of inflection class distinctions (section 2.5).

2.1 Affixes

Affixes are considered the canonical exponent of inflection: they are part of the word form but can be clearly separated from the lexical material. Because of their relative autonomy from the rest of the word form, we can use them to illustrate the various ways that inflection classes can be constituted.

2.1.1 Allomorphy

The most straightforward way for inflection classes to differ from each other is through simple allomorphy. Cross-linguistically, suffixes are the most common type of affix (Himmelmann 2014), so examples of inflection classes based on suffix allomorphy abound; we saw an instance in the Polish paradigms in Table 1.2. Prefixal allomorphy occurs as well, as illustrated in the possessed noun forms in Table 2.1, from Bororo, a Bororoan language of Brazil. (Note that this variation is specific to vowel-initial stems; with consonant initial stems, the prefix-final consonant is deleted and allomorphy is retained only in the third-person singular.)

Table 2.1 *Vowel-initial possessed nouns in Bororo*

	'children'	'food'	'lip'
1SG	it-ore	ik-e	in-ogwa
2SG	ak-ore	ak-e	ak-ogwa
3SG	ore	uk-e	okwa
1INCL	pag-ore	pag-e	pag-ogwa
1PL EXCL	tʃed-ore	tʃeg-e	tʃen-ogwa
2PL	tag-ore	tag-e	tag-ogwa
3PL	et-ore	ek-e	en-ogwa
COREF	t-ore	tig-e	tig-ogwa
RECIP	pug-ore	pug-e	pug-ogwa

Note: Voicing of /k/, as seen in 'lip', regularly occurs under prefixation.
Source: Crowell 1979, Rodrigues 1993.

Here again, we cannot identify an affix for the first-person singular ('my') or plural exclusive ('our'), or third-person plural ('their'), and so on, since the appropriate form depends on the particular noun. Having considered suffixes (Table 1.2) and prefixes (Table 2.1), we should look at infixes, to complete a typology of affixation; however, although we would not exclude the possibility outright, we are not aware of any systems of inflection classes as defined only by infixal allomorphy.

It is worth considering here what lies behind the notion of *allomorphy*. As generally understood, it refers to variation in form that does not have a transparent phonological explanation (though it may well have a phonological condition – see section 5.5.12). Such allomorphy arises in two ways: either affixes which were originally the same diverge phonologically, or originally distinct elements have, through some kind of reanalysis, become equated with each other.

As an example of the first type (phonological divergence), consider the allomorphs of the partitive singular in the Balto-Finnic (Uralic) language Estonian: these are *-t* (/t:/), *-d* (/t/), and Ø, illustrated in (1).

(1) Estonian partitive singular (Rätsep 1982: 56–58)

NOM	PART		reconstructed source
pere	pere-t	'folk'	*pereh-tä
maa	maa-d	'land'	*maa-ta
kala	kala-Ø	'fish'	*kála-a < kála-ða <kála-ta

All three allomorphs go back to a single ancestral suffix *-ta*/*-tä* (subject to vowel harmony), with the consonant subject either to fortition (in the case of consonant clusters, as with 'folk') or lenition (in post-tonic intervocalic position, as with 'fish'), and the final vowel subject to deletion.

Contrast this with the second type (reanalysis of originally distinct elements) as illustrated by the German noun plurals *Tag-e* 'days', *Feld-er* 'fields' and *Ohr-en* 'ears' in (2). The three plural suffixes bear no phonological resemblance to each other (other than the reduced vowel /ə/), and since they occur in the same phonological environment, a phonological explanation does not look promising. But we know the history behind them, which shows that they are three etymologically distinct items. The *-e* suffix goes back ultimately to the Indo-European o-stem thematic nominative plural suffix, whereas the *-er* and *-en* suffixes were originally stem-final elements – their productivity as plural markers in the history of German bearing witness to their reanalysis as suffixes (Wegera & Solms 2000: 1544).

(2) German plurals, and their historical context

			compare Gothic			compare older IE		
SG	PL		SG	PL		SG	PL	
Tag	Tag-e	'day'	dag-s	dag-ōs	'day'	vrk-as	vrk-ās	'wolf' (Sanskrit)
Feld	Feld-er	'field'	riqis	*riqiz-a	'darkness'	tempus	tempor-a	'time' (Latin)
Ohr	Ohr-en	'ear'	ausō	ausōn-a	'ear'	nomen	nomin-a	'name' (Latin)

The different historical origins of allomorphy may ultimately be reflected in aspects of their form (the latter type may show greater phonological divergences among the allomorphs) and their lexical distribution, in that the latter type may mirror some originally meaningful distinction between lexical categories.

2.1.2 *Position*

Affixes are defined not just by their phonological shape, but also by how they fit into the structure of the word form – that is, their position with respect to the stem (Crysmann & Bonami 2016). For example, in the Cushitic language Somali we can say that there is a single set of affixes used by all verbs (Table 2.2). With a small, lexically specified set of verbs they are prefixed, otherwise they follow the root, where /y/ is deleted. (Note that further tense-aspect-mood suffixes may then follow.) Thus, in Somali the inflection classes are defined by the position rather than the shape of the affixes.

We find an even more involved set of structural variants in the possessed nouns of Dano, a Gorokan language of the Trans-New Guinea phylum, spoken in Papua New Guinea. One type involves possessive suffixes, which mark person and number of the possessor, divided into several inflection classes on the basis of allomorphy (Table 2.3).

Table 2.2. *Somali suffix and prefix classes, past simple forms*

	'bring'	'say'
1SG PST	keen-ay	idh-i
2SG PST	keen-t-ay	t-idh-i
3SG M PST	keen-ay	y-idh-i
3SG F PST	keen-t-ay	t-idh-i
1PL PST	keen-n-ay	n-idh-i
2PL PST	keen-t-een	t-idhaah-deen
3PL PST	keen-een	y-idhaahd-een

Source: Saeed 1999.

Table 2.3. *Possessor marking by suffix in Dano*

	'wife'	'younger sister'	'daughter'	'dog'	'head'	'brain'	'friend'
1SG	ele-ʔna	alu-ʔne	alu-ne	gulo-ʔne	godo-ʔne	madu-ʔne	zogo-ne
2SG	ele-nga	alu-na	alu-va	gulo-ka	godo-ka	madu-nga	zogo-ka
3SG	ele-na	alu-no̲lo̲	alu-vo̲	gulo-ʔve	godo-lo̲	madu-no̲	zogo
1PL	ele-de	alu-no̲te	alu-vo̲te	gulo-te	godo-te	madu-de	zogo-le
2PL	ele-di	alu-no̲tine	alu-vo̲tine	gulo-tine	godo-tine	madu-dine	zogo-line
3PL	ele-ʔni	alu-no̲ʔine	alu-vo̲ʔine	gulo-ʔine	godo-ʔnine	madu-ʔnine	zogo-ʔine

Source: Strange 1972.

Another type of noun has possessor prefixes in place of suffixes ('grandchild' in Table 2.4). A third type has suffixes (with a distinct set of allomorphs) plus obligatory suffixes marking case and number of the noun itself ('father'), whereas a fourth type is a combination of the second and third types, with prefixes and the two suffixes ('sister-in-law').

The case suffixes also double as possessor markers, in as much as they occur in two variants: one is for all singular and all first-person possessors, and the other for second-plural and third-plural possessors. We see this basic distinction in Table 2.4 (*-ho* versus *-bo*, for the nominative).

This second set, represented by *-bo* in Table 2.4, shows further differences, as we can see in Table 2.5. Not only do the forms differ, but there is also a failure to distinguish dual and plural of the noun. (Outside of this class, case marking is a phrase-final suffix, and has different allomorphs – for example, *gulo-ʔmo* 'dog-NOM', *gulo-ka ma-ʔmo* 'dog-2SG the-NOM' [= 'your dog'].)

In Dano, then, the inflection class differences are not just a matter of the shape of affixes, but rather they extend to structural differences. And these

Table 2.4. *Other possessor marking types in Dano*

	prefixed 'grandchild'	*suffixed+case* 'father-NOM'	*prefixed+suffixed+case* 'sister-in-law-NOM'
1SG	n-aʔvo	me-ʔne-ho	n-ama-ʔne-ho
2SG	g-aʔvo	m-a-ho	g-am-a-ho
3SG	aʔvo	me-le-ho	ama-le-ho
1PL	l-aʔvo	me-te-ho	l-ama-te-ho
2PL	lang-aʔvo	me-ti-bo	lang-ama-ti-bo
3PL	ng-aʔvo	me-ʔni-bo	ng-ama-ʔni-bo

Note: Underlined vowels are determined by vowel harmony rules.
Source: Strange 1972.

Table 2.5. *Case-number inflection of possessed nouns in Dano*

	'my father'	'my two fathers'	'my fathers'	'your (PL) father'	'your (PL) fathers'
NOM	me-ʔne-ho	me-ʔne-motise	me-ʔne-mote	me-ti-bo	me-ti-nge
ACC	me-ʔne-hine	me-ʔne-motisine	me-ʔne-motine	me-ti-bine	me-ti-ngine
GEN	me-ʔne-hi	me-ʔne-motisi	me-ʔne-moti	me-ti-bi	me-ti-ngi

Source: Strange 1972.

structural differences involve not only affix position, but they involve paradigm shape as well.

Infixes may also play a role in this sort of inflection class opposition; we discuss one such instance from the Daghestanian language Archi in section 5.5.8. Such examples are fairly rare, since infixes (or at least what have been identified as such) are themselves fairly rare compared to prefixes and suffixes.

2.1.3 Distribution

By distribution, we mean the distribution of affixes in the paradigm. Consider the verb paradigms in Table 2.6, from Acazulco Otomí, an Oto-Manguean language of Mexico. This is just a representative fragment of a much larger paradigm. A preposed inflectional formative (*drá* etc.) divides the verbal lexicon into three classes. (The structure of Otomí verb forms and the status of the components of the inflectional formative are matters we return to in section 2.4.) The first class (omitted here) contains distinct allomorphs, but the second and third classes are distinguished from each other only on the basis of the distribution of a suffixed element *-di*, which marks either first and second person, or all three persons, depending on class and aspect-mood value.

Table 2.6. *Acazulco Otomí*

	class II 'walk'		class III 'give'	
REAL INCPL 1	drá	'yǒ=ga	drádí	'ǘn=ga
REAL INCPL 2	grá	'yǒ	grádí	'ǘni
REAL INCPL 3	ra	'yǒ	radí	'ǘni
REAL CPL 1	dídí	'yǒ=ga	dí	'ǘn=ga
REAL CPL 2	gídí	'yǒ	gí	'ǘni
REAL CPL 3	bi	ntx'ǒ	bi	'ǘni
IRR CPL 1	gidi	'yǒ=ga	gidi	'ǘn=ga
IRR CPL 2	gidi	'yǒ	gidi	'ǘni
IRR CPL 3	di	ntx'ǒ	di	'ǘni

Note: *di* suffix underlined.
Source: Hernández-Green 2015.

We could of course treat this as allomorphy, which in a sense it is: for example, the first person realis incompletive is *drá* for class II and *drádí* for class III, and so on. But this would miss the characteristic property of this system. Normally, when we think of inflection class allomorphy we think of one lexeme as having allomorph X and another having allomorph Y. But in this instance, the classes have the same inflectional elements, but they differ in how these elements are distributed in the paradigm. Alternatively, we could treat this as an instance of accidental homophony, with $\textit{-di}_1$ and $\textit{-di}_2$ accounting for the discrepancies in their distribution. But there is comparative evidence that this would be a misrepresentation. Consider the corresponding classes in Tilapa Otomí (Table 2.7), which is morphologically more conservative than Acazulco Otomí. Class III is essentially the same, allowing for the correspondence of Tilapa *-ti* to Acazulco *-di*. But the suffixed element in class II is completely different, consisting of a first person *-du* or *-tu* and a second person *-gu*. Nevertheless, although the forms are different, the paradigmatic *distribution* of the class II suffixes is the same in both Tilapa and Acazulco, that is, it is the same cells where a suffix is present or absent.

There is nothing in the phonological correspondences between the two languages that would suggest that phonological change led to a merger of these suffixes in Acazulco. Rather, it must have been a morphological change. We suggest that what happened was that the class II suffixes were equated with the class III suffixes and so refashioned analogically to match them. We take this as an indication that there is a single suffix whose varying distribution constitutes an inflection class distinction.

Table 2.7. *Tilapa Otomí; correspondences to the -di suffix of Acazulco Otomí underlined*

	class II 'cut wood'		class III 'pray'	
REAL INCPL 1	trá	xu	tráti	xadi
REAL INCPL 2	grá	xu	gráti	xadi
REAL INCPL 3	ra	xu	rati	xadi
REAL CPL 1	túdú	xu	tú	xadi
REAL CPL 2	gúgú	xu	gú	xadi
REAL CPL 3	bi	xu	bi	xadi
IRR CPL 1	gutu	xu	giti	xadi
IRR CPL 2	gugu	xu	giti	xadi
IRR CPL 3	ti	xu	ti	xadi

Source: Palancar 2012.

Table 2.8. *German n-stem nouns*

	SG	PL	SG	PL
NOM	Bär	Bär-en	Frau	Frau-en
ACC/GEN/DAT	Bär-en	Bär-en	Frau	Frau-en

German n-stem ('weak') nouns constitute another example, probably more familiar to most readers (Table 2.8). In this case, the n-stem patterns are just part of a more diverse system of suffixation. Feminine nouns like 'woman' take *-en* in the plural, whereas masculine nouns like 'bear' take this suffix everywhere except the nominative singular.

As pointed out in section 2.1, the *-en* suffix originates in stem morphology. This is probably significant since, as we shall discuss in section 2.2, distributionally defined inflection class distinctions are particularly characteristic of stem alternations, at least under our analysis.

2.2 Stem Alternations

The notion of affixation is reasonably clear; as we have seen, inflectional affixes are elements that can be separated from the lexical material. However, stem alternations are much less clear, since they are made up of everything else which, for whatever reason, the analyst chooses not to segment off the lexical

Table 2.9. *Synopsis of the Wintu verbal paradigm*

indicative stem	citation form	cʔa:wa
	DUB	cʔa:wa-m
	CPL	cʔa:wa-k
	first person	cʔa:wa-da
imperative stem	IMP	cʔa:wu
	SIMUL	cʔa:wu-di
	HORT	cʔa:wu-n
	FUT	cʔa:wu-le
nominal stem	particular NOM	cʔa:wi-t
	particular ACC	cʔa:wi-um
	generic NOM	cʔa:wi-s

Source: Pitkin 1984.

base. Consequently, the types of inflection class systems determined by stem alternations are quite diverse.

At one extreme, stem alternation classes are indistinguishable from inflection classes determined by affix allomorphy. Consider the example of Wintu, a Wintuan language of California. Verbs have three stems, from which a large variety of forms are derived through (possibly multiple) suffixation; a representative sample of these is shown in Table 2.9.

The stems themselves are formed by the suffixation of a theme vowel to the root. A theme vowel, as understood here, is some derivational (originally, at least) element that intervenes between the root and the inflectional affix, and so in a sense is intermediate between a lexical and a grammatical element. The theme vowels vary across the lexicon, dividing it into different inflection classes, whereas the inflectional affixes do not. In effect, we have a three-cell paradigm, whose inflection classes are distinguished by allomorphy of the suffix, illustrated in Table 2.10. There is no difference between this and the sorts of system described in section 2.1.

Thus, although the distinction between stems and inflectional affixes is useful for highlighting the structure of the verb paradigm in Wintu, for the purposes of a typology of inflection classes the distinction is irrelevant: the stem-forming theme vowels can easily be segmented off and treated on par with any other affix. But more often than not, the reason for identifying a particular morphological pattern as a stem alternation is precisely because it is not readily segmentable, so it is treated instead as a modulation of the lexical

Table 2.10. *Wintu theme vowels as inflection classes*[1]

	'sing'	'cauterize'	'be sharp'	'build a house'	'open burrs'	'help'
indicative	cˀa:w-a	bay-a:	cˀic-e:	buq-e:	cˀo:r-a	cˀuq-e:
imperative	cˀa:w-u	bay-u:	cˀic-u:	buq-u:	cˀo:r-u	cˀuq-u
nominal	cˀa:w-i	bay-i	cˀic-e	buq-i	cˀo:r-a	cˀuq-u

	'open nuts'	'preach'	'put away'	'defecate'	'save'
indicative	cˀo:r-a	cˀe:w-a	dukam-a	cˀin-a:	huy-e:
imperative	cˀo:r-u	cˀe:w-u	dukam-e	cˀin-a:	huy-e:
nominal	cˀ:or-o	cˀe:w-e	dukam-a	cˀin-a:	huy-e:

Source: Pitkin 1984.

material (for the issue of segmentation see further in Baerman & Corbett 2012 and Spencer 2012). This of course complicates any discussion of inflectional classes with respect to stem alternations. For affixal material, the notion of classes is relatively unproblematic, because we are dealing with a closed class of items, so that the range of types is correspondingly restricted. But if stem alternations are part of the core lexical material, they are then implicated in the whole gamut of formal variety that characterizes the lexicon, or at any rate the relevant word class.

A fragment of the Georgian verb paradigm is sufficient to demonstrate the sort of structural and phonological variety one can find in stem alternations. The (large) verbal paradigm is based on two stems, examples of which are shown in Table 2.11. These are all transitive active verbs, and all of them take the same set of inflectional suffixes.[2] (The preverbs found in the aorist represent another set of distinctions which cross-cut the ones under discussion here, and will be left aside.) There is no general way to characterize the stem alternations; some verbs appear to have a suffix in the present stem: *-av*, *-am*, *-en*, *-eb*, *-i*, though with one type ('heal') the same element is a fixed part of the verb stem. Other types show a stem-vowel change ('catch'), or insertion ('change'), or no change at all ('forge').

[1] This somewhat reduces the classes Pitkin describes: suffixless forms which show further stem distinctions are omitted, and classes which differ solely in the presence or absence of non-alternating length have been conflated.

[2] Verbs with a present stem suffix *-ob* take *-o* rather than *-a* in the aorist third person singular.

Table 2.11. *Varieties of present ~ aorist stem alternation strategies in Georgian*

	'paint'	'serve'	'pour'	'make'	'heal'
PRS 3SG	xat'av-s	sxam-s	asxen-s	aḳeteb-s	arčeb-s
AOR 3SG	da-xat'-a	daa-sx-a	da- asx-a	ga-aḳet-a	mo-arčeb-a

	'translate'	'change'	'catch'	'forge'
PRS 3SG	targmni-s	cvli-s	č'er-s	č'ed-s
AOR 3SG	gada-targmn-a	še-cval-a	da-č'ar-a	da-č'ed-a

Source: Fähnrich 2012.

Table 2.12. *Macro-classes of the Georgian stem alternations in Table 2.11*

a. suffixing	b. vowel change	c. suffixing + V deletion	d. no change
'paint' (-av) 'serve' (-am) 'pour' (-en) 'make' (-eb) 'translate' (-i)	'catch' (e ~ a)	'change' (-i, Ø ~ a)	'heal' 'forge'

We could certainly construe these all as distinct inflection classes, though their morphological diversity makes it difficult to know what exactly the terms of comparison are. With some degree of abstraction we could treat this as an opposition of classes in structural terms, or macro-classes, which would make the terms of comparison more uniform. Thus, for instance, we could distinguish a class with a suffix alternation (like 'paint'), a class with ablaut (like 'catch'), a class with a suffix alternation and root vowel deletion (like 'change') and a class with no alternations (like 'forge').

And, in fact, this more abstract approach to stem alternations is often the most insightful. Consider the Estonian noun paradigms Table 2.13; these represent a portion of the larger noun paradigm, most of the remaining case forms being derived parasitically from the genitive. First, compare the two stems of 'thought'. Their alternation is characterizable in terms of Estonian's three-way system of foot quantity (where quantity 1 is the shortest and 3 the longest). The stem *mõte* has quantity 2, and the stem *`mõtte* has quantity 3 (graphically

Table 2.13. *Estonian strengthening stem alternation (strong stem shaded)*

	'thought'		'tooth'		'heart'	
	SG	PL	SG	PL	SG	PL
NOM	mõte	\`mõtte-d	hammas	hamba-d	süda	südame-d
PART	mõte-t	\`mõtte-id	hammas-t	hamba-id	süda-t	südame-id
GEN	\`mõtte	mõte-te	hamba	hammas-te	südame	süda-te

Table 2.14. *Estonian weakening stem alternation (strong stem shaded)*

	'bag'		'war'	
	SG	PL	SG	PL
NOM	\`kott	koti-d	sõda	sõja-d
PART	\`kotti	\`kotte	sõda	sõda-sid
GEN	koti	\`kotti-de	sõja	sõda-de

indicated by a tick).[3] The nouns 'tooth' and 'member' have stem alternations which follow the same pattern, but whose phonological realization is different. With 'tooth', both stems are quantity 2, but the stem-medial consonant cluster and the stem-final consonant change. With 'heart', both stems are quantity 1, and differ through a suffixed *-me*. So far this is much the same as in the Georgian examples: there is a phonologically/morphologically diverse family of stem alternations, which we can choose to recognize or not.

When we look at the nouns in Table 2.14 the value of a more abstract treatment becomes apparent. The noun 'bag' has a similar quantity 2 ~ quantity 3 alternation as 'thought' (*koti* ~ *\`kott*(*V*)) in Table 2.13, but its distribution is practically the mirror image. The noun 'war' has the same distribution of stems as 'bag', but in this case both stems are quantity 1, and the stems are distinguished by an alternation in the stem-medial consonant (*d* ~ *j*).

The patterns in Tables 2.13 and 2.14 cover a large portion of the Estonian nominal lexicon. To capture the regularities, we need to take a fairly abstract view. And the grammatical tradition, both of Estonian and of Finnic more generally, recognizes two abstract categories here.

[3] The graphic indication of foot quantity is somewhat involved, and sometimes fails to distinguish between quantity 2 and 3. Therefore, handbooks and grammars often employ the tick across the board to mark quantity 3.

Table 2.15. *Estonian stem classes*

	strengthening	weakening
NOM SG	W	S
PART SG	W	S
GEN SG/NOM PL	S	W
PART PL	S	S
GEN PL	W	S

The first of these is the category of stem gradation. The two alternating stems are characterized as STRONG and WEAK. As the name suggests, strong stems are generally more substantial in some respect, either in a morphophonologically regular way (such as foot quantity or consonant strength), or more idiosyncratically (as with the suffixation in 'heart'). Crucially, this opposition of strong and weak stems applies quite generally across the lexicon, and is largely independent both of paradigmatic patterns and of word classes, since verbs also exhibit the same types of stem alternation. The second category is that of the paradigmatic patterns themselves. These generalize over a phonologically and morphologically diverse range of forms, and are also associated with different case-number suffixes (discussed in more detail in section 6.4). This warrants recognizing them as autonomous morphological structures of their own, conventionally known as the strengthening and weakening patterns (named after the relationship between the nominative and genitive singular stems).

Armed with these two notions of stem gradation and stem pattern, we can represent the contrast between strengthening and weakening patterns as in Table 2.15, that is, as the contrast between two inflection classes.

We must keep bearing in mind that these classes are defined in terms of stem properties, and not actual forms. This means we should be careful about identifying what exactly it is that defines the classes. The labels 'strong' and 'weak' have relevance only relative to each other, and that only within the confines of a single paradigm. They are not associated with any identifiable phonological or morphological property. For example, the nominative singular forms *süda* 'heart' in Table 2.13 and *sõda* 'war' in Table 2.14 are essentially identical in phonological terms, but the first is a weak stem and the second a strong stem. So we cannot make a comparison between forms of different lexemes, only between these abstractions. The key thing about the opposition between these classes is thus not the forms as such, but the pattern of oppositions (strengthening and weakening). That is, these are distributional

Table 2.16. *Portuguese stem alternations*

	'be able'	'fit'	'have'	'play'
1SG IND	posso	caibo	hei	j[ɔ]go
2SG IND	podes	cabes	has	j[ɔ]gas
3SG IND	pode	cabe	ha	j[ɔ]ga
1PL IND	podemos	cabemos	havemos	j[u]gamos
2PL IND	podeis	cabeis	haveis	j[u]gais
3PL IND	podem	cabem	hão	j[ɔ]gam

Source: Maiden 2011.

classes in the sense of section 2.1.3, and not allomorphy-based classes. In this respect they go beyond the type of patterning we observed in Georgian. For an insightful discussion of comparable classes in Sanskrit see Stump (2016: 95–102).

As a rule, inflection classes based on stem alternation will be of the distributional type, precisely because we need to factor out the phonological detail in order to recover any kind of pattern worth talking about. This is even more apparent in systems that lack the kind of morphophonological features represented by strong and weak in Estonian, and consequently lack any ready means of referring to individual stem alternants. For example, Maiden (2011) contrasts two basic stem alternation patterns found in many Romance languages, illustrated here by (rather conservative) examples from Portuguese (Table 2.16). In 'be able' and 'fit', one stem is found in the first singular, another in the remaining forms (the first-singular stem is also found in all the subjunctive forms). The alternation involves either the stem-final consonant ('be able') or stem vowel ('fit').

With 'have' and 'play' there is also stem alternation, this time involving the first and second plural versus the other forms. With 'have' the alternation is realized through the presence or absence of a stem-final consonant /v/; with 'play' there is a stem vowel alternation (not reflected in the orthography, where it is spelled *u* throughout).

Maiden terms these the L-pattern ('be able' and 'fit') and the N-pattern ('have' and 'play').[4] As notions, these patterns correspond to the strengthening and weakening patterns of Estonian, but with the crucial difference that there is

[4] The terms are arbitrary labels, based on the partial resemblance that the alternating patterns have to the dit-dah sequences of the corresponding letters in Morse code. For the persistence of such patterns over time see Maiden (2013).

Table 2.17. *Romance stem patterns (partial)*

	L-pattern	N-pattern
1SG IND	A	A
2SG IND	B	A
3SG IND	B	A
1PL IND	B	B
2PL IND	B	B
3PL IND	B	A

no more general morphophonological pattern such as 'weak' versus 'strong' to allude to. The patterns are defined simply by their oppositions, which we represent in Table 2.17 by A & B: the L-pattern has one stem in the first singular and another one elsewhere, while the N-pattern has one stem in the first plural and second plural and another stem elsewhere (Table 2.17). Even more so than with the Estonian stem alternations, it is obvious here that the opposition is between patterns in the subdivision of the paradigm, and not between forms as such.

Stem alternations can thus be seen as a dimension of inflection class distinctions, but with certain peculiarities determined by their lexical nature. To the extent that we can extract phonologically identifiable elements, they are inflection classes determined by allomorphy (though in such cases one might question how apropos the term *stem alternation* was in the first place). But often it may be more useful to think of stem alternation classes in more abstract terms, either as involving different constructional principles, or different patterns of distribution. A further advantage of this abstract view and further justification for it is that stem alternations and inflections can vary independently of each other: thus the Georgian data (Table 2.13) showed various stem alternations together with identical inflections, whereas the Romance stem patterns (Table 2.19) can occur with different inflections.

2.3 Suprasegmentals

The use of suprasegmental exponents in inflection raises a special set of problems. With segmental exponents we at least stand the chance of isolating inflectional material from lexical material, and in some cases we can even point to a bare stem – that is, a form that lacks any overt inflectional exponents (even here, however, the example of the Aymara accusative in Table 1.1 reminds us to be cautious, since it appears to be formed by truncating the usual final vowel of the stem; in effect, a consonant final noun form is a marked bare

Table 2.18. *Gulmancema tonal patterns*

	'question'	'empty'	'figure out a riddle'	'grill'	'read'	'roll'	'set off on route'	'rinse mouth'
PFV	bágí	báadì	bundì	púlí	cógì	bilnì	bòbí	cùgì
IPFV	bágì	báadí	bundí	púlì	cógí	bilní	bobì	cugì
AOR	bágì	báadí	bundí	púlí	cógì	bilnì	bòbì	cùgì
PFV	HH	HL	L_2L	HH	HL	L_2L	LH	LL
IPFV	HL	HH	L_2H	HL	HH	L_2H	L_2L	L_2L
AOR	HL	HH	L_2H	HH	HL	L_2L	LL	LL

Source: Naba 1994.

stem). But suprasegmentals cannot be effectively separated from the rest of the word form, so their status as lexical or grammatical properties is often obscure. Further, the range of distinctions possible through suprasegmental alternations will often be smaller than that achievable through segmental phonology, so that the number of distinct exponents in such a system will be small. This means that many instances of suprasegmentally determined inflection classes will be defined in terms of distributional differences rather than allomorphy per se.

The verbal paradigms in Table 2.18 from Gulmancema, a Gur language (Niger-Congo) spoken primarily in Burkina Faso, illustrate the major issues. The paradigm consists of three inflected tense-aspect forms. There are three tones: H = high, L = low, and L_2 = low tone which assimilates to the tone of the preceding syllable; the last tone is only distinguishable on initial syllables, since elsewhere there is no context for the assimilation. (Other verbs also employ suffixation, with or without tonal alternations; see section 6.3.) At the very least, we have to assume that every verb has its own inherent lexical tone, since arbitrary tonal contrasts are otherwise a property of the language. For example, the noun *bár-ò* 'king' is high toned and *nàm-ò* 'scorpion' is low toned (both shown here with a low-tone singular suffix), and these are simply lexical properties of the individual nouns.

If we take the first three verbs in Table 2.18, we can provisionally construct a model in which the inflectional tone types are derived from lexical tone: in each of these verbs, the initial tone is unchanging, whereas the final tone switches between the perfective and the imperfective/aorist. If we turn this into a rule of polar tone – for example, 'to form the imperfective and aorist, switch the perfective tone' (or vice versa), then we have no reason to speak of inflection classes. Rather, we have different PHONOLOGICAL classes of verbs

Table 2.19. *Rarámuri stress patterns*

	'dance'	'endure'	'spit'	'pay'
PST	awí-li	anáča-li	kačí-li	natéti-li
COND	awi-sá	anačá-sa	kačí-sa	natéti-sa

Source: Caballero 2011.

with respect to tone, and a single rule of inflection that interacts in predictable ways with this phonology. This is a logically possible system, but it cannot account for the remaining types in Table 2.18. In the case of all but the final verb ('rinse mouth'), the perfective and imperfective exhibit polar tone, but the aorist patterns with the perfective and not the imperfective, whereas in 'rinse mouth', the final syllable displays no tonal change at all. Further, the initial tone in the last two verbs alternates as well. No system of rules will derive the variety of patterns from a single underlying tonal configuration, so there is no way to tease apart lexical and inflectional tone. We thus have inflection classes determined by tone.

The stress system of Rarámuri, a Uto-Aztecan language of Mexico, illustrates a complementary set of issues in suprasegmental inflection classes (the analysis here recapitulates Caballero's (2011) arguments). The conditional and the past tense in Table 2.19 stand for the rest of a (large) verbal paradigm, all of which behaves like one or other of these forms. Consider first the verb 'dance'. In the past tense, the stress falls on the stem-final syllable; in the conditional, it falls on the suffix. One obvious analysis of this would be that certain suffixes, such as *-sa*, attract stress, and that otherwise stress falls on the stem-final syllable. Now compare this to 'endure', which has a trisyllabic rather than a disyllabic stem. The same opposition of stress is seen here, but shifted one syllable towards the beginning of the word. We can maintain the same analysis for this type, with the addition of a phonological rule that adjusts for stem length (the details of which do not concern us here). Now consider disyllabic-stem 'spit' and trisyllabic-stem 'pay'. Here the past tense suffix does not attract stress, which remains on the stem throughout.

Caballero (2011) treats this as the result of the interplay between the accentual properties of stems and suffixes, which can be underlyingly accented or unaccented. (In this sample, 'spit' and 'pay' are underlyingly accented stems, and 'dance' and 'endure' are underlyingly unaccented; *-sa* is an underlyingly accented suffix, and *-li* is underlyingly unaccented.) If the stem is unaccented, stress is assigned by an accented suffix, otherwise it falls on the stem. If the stem is accented, stress always falls on the stem. The principles are outlined in (3).

(3)

	accented stem	*unaccented stem*	
accented suffix	x́-x́ → x́-x	x-x́ → x-x́	+ phonological adjustment based on stem size
unaccented suffix	x́-x → x́-x	x-x → x́-x	

On the face of it, this looks like the kind of system we briefly considered earlier for Gulmancema, and rejected: instead of different inflectional classes, we have different phonological classes of lexemes, and a single set of inflectional markers, which together interact to produce different patterns. But crucially, the phonological property that distinguishes the lexemes – accented versus unaccented – is covert; it has no observable manifestation other than the position of stress itself. Therefore, we have, in effect, two inflection classes of verbs: fixed stress ('accented') and mobile stress ('unaccented'). This is also the sort of system that some have reconstructed for Proto-Indo European, based on similar analyses for Common Slavonic (Olander 2009). In these cases, it has been suggested that the difference between accented and unaccented had its origins in a phonologically overt property such as tone (Dybo 2000).

Alongside the prosodic features of tone and stress, in some languages additional properties of vowels or of the whole syllable play an independent role in inflection and thus can be responsible for inflection class distinctions. For example, in the Mixe languages (belonging to the Mixe-Zoque family of Mexico), vowels are generally understood as being characterized by length, aspiration, and glottalization, which vary independently both of each other and of place of articulation. In Ayutla Mixe, as described by Romero-Mendez (2008), it is these features, and these alone, that divide the verbal lexicon into different inflection classes. These are laid out in Table 2.20, with the five diagnostic forms from which the rest of the verbal paradigm can be deduced. Inflectional exponents are made up partly of a set of prefixes and suffixes that are shared by all verbs, and partly of these vocalic features. Since the affixes induce two morphophonological alternations that affect stem shape, it is worth looking at these briefly so as to be able to factor them out. First, wherever a rule of affixation would yield a sequence /jC/, it surfaces as /Cj/; thus, the prefix *j-* of the dependent forms surfaces after the stem-initial consonant, as in [j-pat-t] → *pjatt* 'ascend.DEP.IRR.3SG'. Second, /j/ causes an umlaut-like alternation (raising and fronting) of a preceding vowel, thus [j-pat-j] → *pjetj* 'ascend.DEP.3SG'.

The vocalic feature alternations, however, are unconditioned, although they are at least in part predictable from stem phonology. But we cannot make any direct association between any of the vowel quality alternations and any of the cells in the paradigm. Consider the breakdown of patterns in Table 2.21.

Table 2.20. *Ayutla Mixe verb classes; numbers refer to the length, aspiration and glottalization classes in Table 2.21*

		'rain'	'write'	'saw'	'laugh'	'arrive'	'sprain foot'	'ascend'	'speak'	rest'	'come'	'buy'	'hear'	'carve'	'collapse'
		CVˀV	CVˀVy	CVVC	CVˀVC	CVˀVC	CVX	CVC	CVʰCʂ	CVˀCʂ	CVN	CVj	CVV	CVVh	CVˀV(j)
IMP	(Ø)	tuˀu	hʌˀʌj	hɨɨt	ʂeˀek	hʌˀʌt	haʂ	pat	kʌʰpʂ	poˀkʂ	men	hɨj	mɨtoo	pʌʌh	hʌˀʌ
3SG	(-p)	tuup	haapj	hɨˀtp	ʂekp	hʌˀtp	haʂp	paʰtp	kʌʰpʂp	poˀkʂp	memp	hɨɨpj	mɨtoop	pʌʌhp	hʌˀʌp
DEP 3SG	(j-…-j)	tjuˀuj	hjʌˀʌj	hɨɨtj	ʂjiˀikj	hjʌˀʌtj	hjeʂj	pjetj	kjaʰpʂj	pjeˀkʂj	mjinj	hɨj	mjɨtej	pjʌʌhj	hjʌˀʌj
DEP IRR 3SG	(j-…-t)	tjuˀut	hjʌˀʌtj	hɨɨtt	ʂjeˀekt	hjʌˀʌtt	hjaʂt	pjatt	kjʌʰpʂt	pjoˀkʂt	mjeˀnt	hɨˀɨtj	mjɨtoˀot	pjʌʌht	hjʌˀʌt
CPL 3SG	(Ø)	tuu	hʌʌʰj	hɨˀɨʰt	ʂeeʰk	hʌˀʌt	haʰʂ	paaʰt	kʌʌʰpʂ	pooˀkʂ	meen	hɨɨj	mɨtoo	pʌʌʰh	hʌˀʌ
length				2	2	2		4	4	4	4	3	1		
aspiration			2	2	2		2	1						2	
glottalization		3	3	2	3						1	1	1		

Note: Where affixation would yield a sequence /jC/, it surfaces as /Cj/; /j/ causes umlaut of preceding V.
Source: Romero-Mendez 2008.

Table 2.21 *Key to alternation patterns in Table 2.20*

	length [VV]				aspiration [h]		glottalization [Vˀ]		
	class 1	class 2	class 3	class 4	class 1	class 2	class 1	class 2	class 3
IMP	■	■							■
3SG	■				■			■	
DEP 3SG		■							■
DEP IRR 3SG	■	■	■				■		■
CPL 3SG	■	■	■	■	■	■			

Note: Shading registers the presence of the prosodic exponent, according to different paradigmatic patterns for each class.

In the independent neutral third singular, for example, some verbs shorten their stem vowel, whereas others lengthen it. Some verbs add glottalization, whereas some lose it. As with the Gulmancema tone example in Table 2.18, we cannot derive the vocalic feature alternations from the interaction of the stem with an invariant set of floating features. Rather, we have different inflection classes whose only manifestation is through variation in the patterning of these features in the paradigm. (Equally, the patterns are not wholly predictable from stem phonology: compare 'laugh' and 'arrive', which share the same stem shape; 'arrive' has fixed glottalization throughout the paradigm, but with 'laugh' it alternates.)

2.4 Uninflectedness

So far we have looked at differences in the form and patterning of inflectional marking across lexemes. But we should not forget that whether or not a lexeme inflects in the first place may itself be a non-trivial morphological property. This is a feature of the inflection class system of Russian nouns. There are six indisputable case values (the remaining, very interesting values are discussed in detail in Corbett 2012: 200–222), distinguished in both singular and plural. As we show later (section 5.2), there are four main inflection classes. In Table 2.22 we present an instance of class II, and paradoxically three instances of "class V":

The first noun, *kniga* 'book,' shows why we need two number values and six case values for the inflection of Russian nouns. However, *pal′to* 'coat' and *kenguru* 'kangaroo', and numerous similar nouns, show no sign of inflection. Yet they are not defective. They appear in all syntactic environments appropriate for nouns. If they require agreement, then agreement targets agree

Table 2.22 *Inflectional classes II and V in Russian*

	Class II	Class V	Class V	Class V (defective)
	'book'	'coat'	'kangaroo'	'riding breeches'
NOM SG	knig-a	pal´to	kenguru	–
ACC SG	knig-u	pal´to	kenguru	–
GEN SG	knig-i	pal´to	kenguru	–
DAT SG	knig-e	pal´to	kenguru	–
LOC SG	knig-e	pal´to	kenguru	–
INS SG	knig-oj	pal´to	kenguru	–
NOM PL	knig-i	pal´to	kenguru	galife
ACC PL	knig-i	pal´to	kenguru	galife
GEN PL	knig	pal´to	kenguru	galife
DAT PL	knig-am	pal´to	kenguru	galife
LOC PL	knig-ax	pal´to	kenguru	galife
INS PL	knig-ami	pal´to	kenguru	galife

appropriately, but these nouns can equally well appear in syntactic positions where there is no overt indication of their number and case. We may treat them as another inflection class, as in Corbett & Fraser (2000: 308), or treat them as somehow different, but the key point is that their inflectional behaviour, or lack of it, has no repercussions outside the morphology.

Nouns of this type are also rather like the other inflection classes in their behaviour according to other properties. For instance, though they cannot show the characteristic syncretisms induced by animacy (for which see Table 6.3), they do require the appropriate agreements, with *pal´to* 'coat' being inanimate and *kenguru* 'kangaroo' being animate. And then the final example in Table 2.22 has an interesting relation with number. The noun *galife* 'riding breeches' does not inflect but is also a *plurale tantum* noun, as shown by the fact that it takes only plural agreements (Isačenko 1962: 77). It is therefore defective as well as non-inflecting. Here is an instance:

(4) v ser-yx galife
in grey-PL.LOC riding breeches
'in grey riding breeches' (From Ju. Trifonov, *Dom na naberežnoj*)

This fits into a pattern, in that several other nouns denoting various types of paired object behave similarly (see Corbett 2000: 175–176 for discussion). However, these nouns are nonetheless defective, in that there are contexts for which they have no appropriate form, and where a repair strategy is required.

Table 2.23. *Polish 'museum'*

	SG	PL
NOM	muzeum	muzea
ACC	muzeum	muzea
GEN	muzeum	muzeów
DAT	muzeum	muzeom
LOC	muzeum	muzeach
INS	muzeum	muzeami

We might assume that there was a single structural issue here – that items could inflect or fail to inflect. The situation is more complex: there is an interesting (but very small) set of lexemes in Polish which are split by inflectability. This was pointed out by Laskowski (1990), and the nouns are listed in Swan (2002: 118). An example is given in Table 2.23.

Such nouns have a split between the non-inflecting (singular) and inflecting (plural) subparadigms. This demonstrates that (non-)inflectability is not necessarily a property of whole lexemes, but that it can affect parts of the paradigm of individual lexemes. Equally, it can affect the marking of a part of the feature set. Recall the Dano possessed nouns illustrated in Tables 2.3–2.5. One group of nouns inflects for case and another does not; therefore we can say that this second group is uninflected for case. But it is still inflected for possessor person and number. Some nouns inflect for possessor person and number, and other nouns add case inflection to that.

2.5 Words as Inflection Class

Inflection classes are manifested over inflectional morphology, by definition of course. But on occasion we find words which themselves are distributed according to principles like those that govern inflection classes. The sort of thing that might spring to mind is auxiliary selection in the Romance and Germanic languages, or in Basque. In these languages verbs choose between different auxiliaries – for example, a 'be' auxiliary and a 'have' auxiliary: thus German *ich bin gelaufen* 'I have walked' (lit. 'I am walked') vs. *ich habe gegessen* 'I have eaten' (lit. 'I have eaten'); in some Italian dialects there are even finer-grained distinctions, manifested morphologically by various mixtures of the two auxiliaries determined by subject person and number (Loporcaro 2007). In practice, the favoured explanation for this variation in auxiliary choice is not arbitrary lexical specification. Rather, the distinction is thought to reflect syntactic properties such as transitivity or reflexivity, and semantic properties such as active~inactive or unergative~unaccusative, or resultativeness. This is in keeping with standard assumptions about the architecture of grammar, in which arbitrary morphological properties do not play a role in syntactic rules (Miller, Pullum, & Zwicky 1997). That said, it may be wise to be circumspect about claims of semantic transparency, and leave open the possibility that at least some portion of auxiliary choice is lexically determined.

A more unequivocal example of lexical specification comes from the Otomí languages, here exemplified by Tilapa Otomí (also discussed in section 2.1.3), though it is less clear to what degree the element in question is really a separate word. Finite verbs consist of two parts: a lexical stem, preceded by a formative that marks tense-mood and subject person. The order of these two parts with respect to each other is fixed, and nothing can intervene, but otherwise the phonological connection between the two is weak. Although the preverbal formative typically procliticizes to the verb, under the right conditions it can encliticize to a preceding word, allowing for a break between it and the lexical verb, as illustrated by the pauses (#) and readjustments shown in (5), taken from Palancar & Baerman (2014). Therefore, it seems fair to say that the preverbal element is a clitic, giving it the status of a separate but prosodically dependent and syntactically restricted word.

(5) a. Hesitation pause
pws grá=nde gi=^{h}pöh=wi
well 2.INCPL.REAL=want 2.CPL.IRR=know=PL
a te=**tú#** *sufre*=ga=‘mbe
DEF.SG what=1.CPL.REAL suffer=1=PL.EXCL
‘Well, if you want to know what we suffered.’ (Txt, Tilapa)

b. Pause & repair
ha̲n=**dú#...** ha̲m=**bi** hih-ki
again=1.CPL.REAL again=3.CPL.REAL bathe-1.OBJ
‘And again…again she bathed me.’ (Txt, Tilapa)

This inflectional clitic falls into three classes, depending on the lexical verb it is associated with. Table 2.24 gives an extract of the larger paradigm, illustrating the ways in which the forms and classes differ.

Unlike the familiar European examples alluded to earlier, this grouping of lexemes does not follow any discernible syntactic or semantic partitioning of the lexicon. A certain syntactic correlation does exist: class II consists exclusively of intransitive verbs, as shown in Table 2.25, derived from a corpus of 630 verbs. The other two classes, however, are evenly mixed between transitives and intransitives, and no further semantic correlation has yet been found.

Table 2.24. *Three classes of clitic formative in Tilapa Otomí. Formatives unique to a given class are shaded*

			class I 'water'		class II 'cut up wood'		class III 'pray'	
REALIS	INCPL	1	trá	xithe	trá	xu	trá tí	x<u>a</u>di
		2	grá	xithe	grá	xu	grá tí	x<u>a</u>di
		3	ra	xithe	ra	xu	ra ti	x<u>a</u>di
	CPL	1	tú	xithe	tú dú	xu	tú	x<u>a</u>di
		2	gú	xithe	gú gú	xu	gú	x<u>a</u>di
		3	bi	xithe	bi	xu	bi	x<u>a</u>di
	PRF	1	xtú	xithe	xtú dú	xu	xtú	x<u>a</u>di
		2	xkú	xithe	xkú gú	xu	xkú	x<u>a</u>di
		3	xú -n	xithe	xpi	xu	xpi	x<u>a</u>di
	PPRF	1	xtá	xithe	xtá dú	xu	xtá	x<u>a</u>di
		2	xkí	xithe	xká gú	xu	xkí	x<u>a</u>di
		3	xká -n	xithe	xká	xu	xkí	x<u>a</u>di
IRREALIS	INCPL	1	gra	xithe	gra tu	xu	gra ti	x<u>a</u>di
		2	gra	xithe	gra gu	xu	gra ti	x<u>a</u>di
		3	tra	xithe	tra	xu	tra ti	x<u>a</u>di
	CPL	1	gu	xithe	gu tu	xu	gi ti	x<u>a</u>di
		2	gi	xithe	gu gu	xu	gi ti	x<u>a</u>di
		3	ta	xithe	ti	xu	ti	x<u>a</u>di
	PST	1	gugu	xithe	gugu tu	xu	gugu	x<u>a</u>di
		2	gigi	xithe	gu gu	xu	gigi	x<u>a</u>di
		3	tigi	xithe	tigi	xu	tigi	x<u>a</u>di
	PRF	1	xkugu	xithe	xtigu tu	xu	xkugu	x<u>a</u>di
		2	xkigi	xithe	xtigu gu	xu	xkigi	x<u>a</u>di
		3	xti	xithe	xtigi	xu	xtigi	x<u>a</u>di
	PROSP	1	xtágu	xithe	xtágu tu	xu	xtágu	x<u>a</u>di
		2	xtági	xithe	xtági	xu	xtági	x<u>a</u>di
		3	xta	xithe	xtagi	xu	xtagi	x<u>a</u>di

Source: Palancar 2012.

Table 2.25. *Distribution of inflection classes by transitivity in Tilapa Otomí*

	class I	class II	class III
transitive	204	0	87
intransitive	191	72	76

Source: Palancar 2012.

This means that we cannot understand these preverbal formatives as functionally opposed auxiliaries. At most we can say that the repertoire of inflection classes is conditioned by transitivity: intransitive verbs are spread over three inflection classes, whereas transitive verbs are spread over just two.

These formatives do not resemble auxiliaries in any other respect. They are not viable words on their own. Morphologically, there is no consistent distinction between the classes, such as one might expect from auxiliaries – for example, an opposition of distinct roots. Rather, the core of the forms is largely the same, differing mainly through variably distributed suffixed elements. Thus, class I has a terminal *-n* in some third person forms, or lacks a terminal *-gi* where the other two have it; class II has a terminal *-tu* (or *-du*)/*-gu* pairing in many of the first and second person forms, and class III typically has a terminal *-ti*; there are also some vocalic differences between the classes as a well. All told, the inflection classes defined by the preverbal clitics resemble those we might expect from affixal morphology, were it not for the potential prosodic gap that separates them from the rest of the verb.

The Finisterre-Huon languages of the Trans-New Guinea phylum provide further evidence of the possible parallels between lexical selection and inflection classes. In this group of languages, transitive verbs take object suffixes that intervene between the stem and the subject person-number suffixes. The three classes of Selepet in Table 2.26 are a representative example.

These object markers correspond to independent verbs (Table 2.27). The class I suffixes correspond to 'see', class II corresponds to 'bite' or 'give' (the two senses are distinguished in the third singular, where 'give' has a suppletive stem), class III corresponds to 'hit/kill'. The one deviation between the object markers and the free verbs is in the third singular, where classes I and II lack a suffix, even though there is a corresponding form in the paradigm of the free verb.

Presumably the object markers were derived by incorporation of the free verb forms. But the principles of selection are obscure; for example, in the

Table 2.26. *Object-marking classes in Selepet*

	I 'cut'	II 'await'	III 'touch'
1SG	gâl-nek-	mabot-nihi-	walip-noho-
2SG	gâl-gek-	mabot-gihi-	walip-goho-
3SG	gâl-	mabot-	walip-ku-
1DU	gâl-nelek-	mabot-nitki-	walip-notko-
2/3DU	gâl-yelek-	mabot-yitki-	walip-yotko-
1PL	gâl-nenek-	mabot-ningi-	walip-nongo-
2/3PL	gâl-yek-	mabot-yingi-	walip-yongo-

Source: McElhanon 1972.

Table 2.27. *Object verbs in Selepet*

	'see'	'bite'	'give'	'hit, kill'
1SG	nek-	nihi-	nihi-	noho-
2SG	gek-	gihi-	gihi-	goho-
3SG	ek-	ihi-	waŋ-	ku-
1DU	nelek-	nitki-	nitki-	notko-
2/3DU	yelek-	yitki-	yitki-	yotko-
1PL	nenek-	ningi-	ningi-	nongo-
2/3PL	yek-	yingi-	yingi-	yongo-

Source: McElhanon 1972.

case of Ono, a related language of the Finisterre-Huon family, Wacke (1930–1931: 177–179) suggests that the 'see' verb is used as a direct object marker, and that 'give' is used as an indirect object marker. But in discussing Selepet, Foley (1986: 141) observes that 'we are forced to conclude that the selection is basically arbitrary, although in some restricted areas semantic principles may be at work'. This suggests that the principles of lexical selection on the one hand, and inflection class assignment on the other, are not necessarily so different, even though one plays out in the realm of syntax and the other in morphology.

2.6 Conclusion

Any defining property of an inflectional paradigm – be it phonology, morphotactics or the very shape of the paradigm – may be subject to variation across lexemes, and hence, be a domain of inflection class distinctions. Recognizing this is in no way a denial of qualitative differences between systems. For example, the potential for generating a multitude of inflection classes via allomorphy is far greater for affixal than for prosodic exponents, since affixes typically have recourse to a greater range of phonological distinctions. On the other hand, affixes only rarely show the kind of paradigmatic variation responsible for distributional classes, so this category largely applies to stem alternations whose phonological details have been abstracted out. And of course, this big tent view of inflection classes may be too expansive for certain purposes, particularly as some morphological models may apportion different exponent types to different portions of the grammar. For example, Carstairs-McCarthy's well-known proposal about constraints on the proliferation of inflection classes (originally the Paradigm Economy Principle, later modified as No-Blur and Vocabular Clarity; see Carstairs 1983, Carstairs-McCarthy 1994, 2010) expressly applies only to affixally determined inflection classes. The possible typological variety in the manifestation of inflectional differences across lexemes makes it all the more important to define precisely what one does and does not mean by the term *inflection class*.

3 *Features*

It seems reasonable that any morphosyntactic feature that defines a paradigm should also be subject to inflection class distinctions. But neither should we assume unquestioningly that they are. We survey inflection class systems here according to several cross-linguistically common morphosyntactic features: case (section 3.1), number (section 3.2), person (section 3.3), gender (section 3.4), and tense/aspect/mood (section 3.5). The examples have been selected so that they are pure, that is, where we can isolate the morphological marking of that feature and thereby know for sure that that is the feature whose realization is bringing about the inflection class distinctions.

3.1 Case

In Bauzi, an East Geelvink Bay family language of New Guinea, case is the only feature marked on nouns, dividing them into different classes based on allomorphy of the case marker (Table 3.1). This example is striking through the absence of number marking, since most familiar examples of inflection classes, which are due to case marking (e.g., in Indo-European or Uralic), involve the cumulative marking of number as well. In Bauzi, number plays a morphological role only with verbs, some of which have a stem alternation or suppletion according to the number of the absolutive argument.

3.2 Number

Lexical variation in nominal number marking is a theme that runs through many languages, and accounts for the most extreme proliferation of allomorphy and inflectional variation, so much so that it is worth dwelling on it here in some detail. Above all, this is due to inflectional variety in plural marking. For example, Nuuhchahnulth, a Wakashan language of British Columbia (Canada), employs three very different devices to mark noun plurals: suffixation (*-m̓inḥ* is productive), infixation of <*t*>, and reduplication (Table 3.2).

Table 3.1. *Bauzi case inflection*

	'banana'	'pigeon'	'canoe'	'cassowary'	'bushknife'
ABS	fem	mek	ih-e	bih-i	ae
ERG	fem-ot	mek-it	ih-at	bih-et	ae-t
DAT	fem-o	mek-i	ih-a	bih-e	a<l>e

Source: Briley 1997.

Table 3.2. *Number marking in Nuuhchahnulth nouns*

	'loon'	'canoe'	'man'	'fish'	'house'
SG	ʔa:ma	č̓apac	čakup	haʔum	maḥt̓i:
PL	ʔa:ma-m̓inḥ	č̓ay̓a:p-ac	ča:kup-i:ḥ	ha:<t>ʔum	<u>ma:</u>maḥt̓i:

Source: Davidson 2002.

Table 3.3. *Number marking in Lavukaleve nouns*

	'towel'	'shoe'	'shoulder'	'plantation'
SG	taol	buti	fakas	planteson
DU	taol-al	buti	fakas-al	planteson
PL	taol-kal	buti-kal	fakas-aul	planteson-aul

	'rafter'	'forehead'	'bell'	'moon'	'eye of greensnail shell'	'war'
SG	hoho	hai	belo	kua	kua	wo
DU	hoho-l	hai	belo-l	kua	kua-ol	wo
PL	hoho-n	hai-n	belo-vil	kua-vil	kua	wo

Source: Terrill 2003.

In Lavukaleve, of the Central Solomons family of New Guinea, a wide variety of different plural markers is supplemented by dual markers, whose variety is much more restricted (Table 3.3). But whether a noun actually uses a dual marker is itself a matter for lexical specification. Where the dual marker is absent, the dual form is identical to the singular – namely, the bare stem (but the noun can still take dual demonstratives, for example); compare the pairs of nouns in Table 3.3, where one noun has the dual marker and the other does not. Thus, the mere presence or absence of distinct dual marking potentially doubles the number of inflection classes.

Table 3.4. *Number marking in !Xoon nouns*

	'fillet'	'black crow'	'eyelash'	'(type of) lizard'	'woman's rear apron'	'wild cotton'	'hunting bag'
SG	ǁɢǫ́lo	ǁnòha-là	gǀqhùbi-sà	ǁɢà̰hni-kà	ǁgábi-tê	!gúu-bè	!ūma
PL	ǁɢǫ́lo-tê	ǁnòha-tê	gǀqhùbi-tê	ǁɢà̰hni-tê	ǁgábu-tê	!gúu-tê	!ūma-nî

Source: Traill 1994.

A much rarer configuration is for the bulk of inflection class distinctions to be defined by allomorphy in the singular. For example, in !Xoon, a Tuu language of Botswana, nearly all nouns have the same plural suffix *-tê*, but a wide variety of singular suffixes is found (Table 3.4). (Note that 'woman's rear apron' takes the 'plural' suffix in the singular as well, and marks number through an alternation of the stem-final vowel.) That said, other plural suffixes do occur, such as *-nî*, so it is not the perfect mirror image to a language like Nuuhchahnulth, in which class distinctions emerge entirely from the shape of plural marking.

Variation in singular marking more typically runs parallel to variation in non-singular marking, nowhere more characteristically than in the sorts of systems that have been dubbed 'tripartite' number systems, in reference to Nilo-Saharan languages (Dimmendaal 2000). But these systems are not limited to this family. The paradigms in Table 3.5, from the Zaparoan language Arabela, spoken in Peru, are equally representative. Some nouns take an overt plural suffix, and some nouns an overt singular suffix. Both of these have numerous allomorphs. Still other nouns take both singular and plural suffixes – hence some use the term tripartite number, in that there are three schemes of number marking. The lexical distribution of these types is suggestive of some kind of semantic logic: in particular, the nouns with a morphologically marked singular can typically be understood as having a collective or mass sense (e.g., insects, palm fiber); where only the singular is morphologically marked, this is sometimes known as a *singulative*. The same semantic properties also seem to characterize nouns whose singular and plural are both overtly marked, except, in this case, the collective (= plural) form is not basic but is formed by means of a suffix.

Although the number markers in Arabela appear to be fixed to either a singular/singulative or a plural/collective function, in Murle, a Surmic language (Nilo-Saharan) of South Sudan, at least some of the number markers lack a dedicated function. Thus the nouns in Table 3.6a display the three different suffixes *-ɛn*, *-it* and *-t* that function either as marked plural or marked singular.

Table 3.5. *Number marking in Arabela nouns*

a. plural suffixes					
	'(type of) bird'	'wing'	'girl'	'house'	'muscle'
SG	suunu	nesu	niyacoo	tia	shiyonu
PL	suunu-pue	nesu-hua	niyacoo-jori	tia-ca	shiyonu-na
b. singular suffixes					
	'mosquito'	'moth'	'(type of) palm'	'(type of) palm fiber'	
SG	naa-su	susu-nu	sho-naja	shiquia-mue	
PL	naa	susu	sho	shiquia	
c. singular and plural suffixes					
	'black mushroom'	'fruit'	'group'	'thing'	
SG	surususu-ru	nosu-taca	maa-nu	casa-a	
PL	surususu-hua	nosu-na	maa-pue	casa-mi	

Source: Rich 1999.

Table 3.6. *'Marked singular' and 'marked plural' in Murle*

a. singular or plural suffixes						
	'womb'	'army'	'small gourd'	'fly'	'lion'	'goat'
SG	mom	diic-ɛn	diic	kiroŋ-it	maa	bawo-t
PL	mom-ɛn	diic	diic-it	kɛroŋ	maa-t	bawo
b. singular and plural suffixes						
	'bellows'	'thorn'	'chief'	'tree'		
SG	akub-a	bil-ɛt	alaa-n	kɛɛ-t		
PL	akub-ɛt	bil-a	alaa-t	kɛɛ-n		

Source: Arensen 1982.

Equally, in paradigms in which both numbers are overtly expressed, they can appear in two configurations, as shown in Table 3.6b. (In both sets the suffix *-t* could be interpreted as a post-vocalic variant of *-it* or *-ɛt*.) In such a system, alongside inflection class distinctions due to simple allomorphy (and Murle has quite a bit of that), we have differences defined by the paradigmatic distribution of allomorphs.

Table 3.7. *Plural allomorphy in nouns and verbs (singular event forms) in Seri*

nouns			verbs		
SG	PL		SG SBJ	PL SBJ	
kó:pa	kó:pa-x	'drinking glass'	-askíta	-askíta-x	'refuse to share'
ʔaχ	ʔáχa-xam	'fresh water'	-okósi	-okósi-xam	'bite and suck'
isɬí:k	isɬí:k-xox	'left hand/arm'	-aʔí:ʔom	-aʔí:ʔom-xox	'ambush'
i:χ	i:x-k	'water'	-apoá:χ	-apoá:x-k	'lean'
ʃa:x	ʃá:ɬ-ka	'cave'	-iʔínex	-iʔíneɬ-ka	'be exposed'
isɬá::x	isɬá:iʃi-ɬ	'(his/her) shoulder'	-atá:ix	-atá:iʃi-ɬ	'spin (a top)'
naxó:	naxó:-ɬka	'spotted sand bass'	-sítoχ	-sítoχ-ɬka	'strike (a match)'
ʔakáʃoɬ	ʔakáʃoɬ-kam	'piece of cloth'	-a:kx	-a<tó:>kaɬ-kam	'do twice'
i:f	í:f-kox	'nose, beak'	-é:fe	-é:f-kox	'be hook shaped'
kset	ksét-ikoɬ	'magnificent frigatebird'	-kes	-kés-ikoɬ	'cover completely'
ʔéʔe	ʔéʔe-t	'plant'	-a	-a-t	'come'
i:χʷt	í:χot-a	'sea current'	-akésχ	-akéseʃχ-a	'distribute to everyone'
kanóa:	kanóa:-tax	'boat'	-aχá:ʃa	-aχá:ʃa-tax	'drip (something thick)'
o:iɬ	ó:iɬ-o	'(its) crest'	-o:nɬ	-ó:nɬ-o	'stir'
χtá:si	χtá:si-tox	'estuary'	-éme	-éme-tox	'run out'
nop	nóp-χam	'bobcat'	-sap	-sáp-χam	'swat at'

Note: The verb forms here have been stripped of their prefixes, which mark subject person (and, for first person, number), tense-mood, object person, and negation; prefixation shows no allomorphy which is not phonologically determined.

Source: Moser & Marlett 2010.

Of course, number is not a feature just of nouns. Among other things, it shows up in agreement, often realized cumulatively with other features such as case, gender, person, or tense, aspect and mood (TAM). For a relatively 'pure' example of number-marking allomorphy that extends outside the nominal system, Seri, a language isolate of Mexico, offers a particularly striking example. Nouns show a dizzying array of different plural suffixes, nearly all of which are also found marking plural subjects in verbs (Table 3.7). These forms also show various types of stem alternation which are also shared across the two word classes – for example, the stem-final consonant alternation in *ʃa:x* ~ *ʃá:ɬ-ka* 'cave' alongside *-iʔínex* ~ *-iʔíneɬ-ka* 'be exposed'.

Table 3.8. *Allomorphy of event plurality marking in Seri (plural subject forms)*

	'inhale'	'bounce'	'carry'	'hurry to do something carelessly'
SG EVENT	-xaχ-tox	-aʔaːʔnaɬ-ka	-óːne-k	-anámɬ-kox
PL EVENT	-xaχa-xam	-aʔaːʔnaɬ-kox	-óːne-xox	-anámɬ-kam

Source: Moser & Marlett 2010.

Table 3.9. *Subject person marking in Jamul Tiipay*

	'run'	'sing'	'be big'	'be fat'
1	naw	cheyaw	aʔ-tay	aʔ-shay
2	me-naw	me-cheyaw	me-tay	me-shay
3	we-naw	cheyaw	kwe-tay	we-shay

Source: Miller 2001.

In fact, the plural allomorphs have an even wider application than is shown in Table 3.7. Alongside subject number, Seri verbs make a cross-cutting distinction in event number. The forms in Table 3.7 are specific to singular events. The marking of plural events draws heavily from the same stock of morphological devices as plural subjects and, consequently, shares its diversity of allomorphy. The forms in Table 3.8 have the same plural suffixes as found in Table 3.7, variously deployed to mark event number. Thus, in Seri, plural marking – and the inflection classes that result from its allomorphy – is to some extent indifferent about whether this plurality is proper to nouns or verbs and whether it is an inherent feature or one of agreement. (This system is discussed in more detail in section 6.2.1.)

3.3 Person

Inflection for person is often cross-categorical, marking arguments on verbs and possessors. The paradigms in Table 3.9, from Jamul Tiipay, a Cochimí-Yuman language of Baja California, illustrate inflection classes determined by the marking of subject person on verbs, with classes contrasting both through the presence or absence of affixes, and through allomorphy (third person *kwe-* vs. *we-*). Subject number is an independent category – morphologically quite diverse, typically involving stem alternations of one kind or another, and affecting only a subset of verbs (and even with these it is not obligatory).

Table 3.10. *Possessor person marking in Karajá*

	'be crazy'	'be angry'	'forehead'	'hand'	'blood'
1	wa-itʃẽʤɛ=re	wad-ɛbʊrɛ=re	wa-kɔrʊ	wad-ɛbɔ	wal-əbʊ
2	a-itʃẽʤɛ=ʤe	ɛbʊrɛ=ʤe	a-kɔrʊ	ɛbɔ	əbʊ
3	i-itʃẽʤɛ=re	ʤ-ɛbʊrɛ=re	i-kɔrʊ	ʤ-ɛbɔ	ʤ-əbʊ

Source: Ribeiro 2012.

Table 3.11. *Possessor person marking on kinterms in Kobon*

	'brother'	'maternal aunt'		'same sex cousin'		'elder sibling'
1	mam	amañ		bane	(< ba-ane)	mədap
2	na-mam	namañ	(< na-aman)	nane	(< na-ane)	na-mdaŋ
3	nə-mam	nɨmañ	(< nə-aman)	nɨne	(< nə-ane)	nɨ-mdaŋ

Source: Davies 1981.

The paradigms in Table 3.10, from Karajá, a Je language of Brazil, illustrate inflection classes in terms of possessor person. In this language, any noun can take possessive inflection, and such a form is also used for descriptive predicates, which are morphological nouns that serve the purpose of predicate adjectives or stative verbs. Both 'ordinary' nouns and descriptive nouns (here shown with the imperfective enclitic, which is also sensitive to person) are split into two classes on the basis of the third-person prefix; the 'ordinary' nouns, here represented by body parts, are further split on the basis of the first-person prefix. Note that in Karajá, number is not an inflectional category, so these are all pure markers of person.

Although the Karajá classes of possessor person obtain quite generally across the nominal lexicon, it is often the case that inflectional differences are restricted to a closed set of nouns such as kin terms or body parts, the sort sometimes known as inalienably possessed. The examples in Table 3.11, from Kobon, a Trans-New Guinea language, are a typical case: morphological marking of possessor person (number is not distinguished) is restricted to kin terms.[1] 'Brother' and 'maternal aunt' represent the regular case (such as it is, given the small number of possessed nouns), with no prefix with first-person possessors, *na-* with second person and *nə-* with third person (there is some

[1] Here again we see the issue of motivation. 'Inalienably possessed' is a recurring characterization of sets of nouns, which match a semantic class to a greater or lesser extent. These nouns inflect for feature values of the possessor, remaining nouns do not. Contrast that with Russian (section 2.4) in which the set of uninflected nouns cannot be characterized in semantic terms, neither positively nor negatively.

Table 3.12. *Possessor person marking in Golin*

	type 1		type 2	
	'son'	'buttocks'	'friend	'chest'
1	wan-an	den-an	en-an	din-an
2	wan-in	den-in	en-in	din-in
3	wan	den	en-in	din-in

Source: Evans, Besold, Stoakes, & Lee 2005; see also Bunn 1974.

coalescence between prefix and stem, noted in the table). The other two types shown here differ in their first person forms: 'same sex cousin' has a prefixed *ba-*, whereas 'elder sibling' has a suppletive form.

In Golin, also a Trans-New Guinea language, the inflection classes are distinguished distributionally, rather than by allomorphy (Table 3.12). With some nouns the suffix *-in* is used just for second-person possessors (type 1), with others it is used for the third person as well (type 2), yielding an inflection class distinction of the distributional type. There is no obvious semantic conditioning factor (e.g., the classes cross-cut animacy). Phonologically, there is an asymmetry between the stems of the two classes: type 1 consists entirely of nouns with stems ending in /n/, whereas type 2 appears to be an open class, but also includes nouns with stems in /n/.

3.4 Gender

We are not aware of any systems in which inflection class distinctions are manifested by the marking of gender in isolation. All the instances we know of involve gender marked together with some other feature, typically number or person, or, in the Indo-European languages, case as well. It is not clear whether we should attribute any significance to this gap, but we suspect not; cross-linguistically, gender is not typically 'pure', so the gap may simply be due to the infrequency of 'pure' gender marking.

3.5 Tense/Aspect/Mood

Inflection classes determined solely by aspect/mood marking are found in Zenzontepec Chatino, an Oto-Manguean language of Mexico, where prefixes and their combinations occur in several variants. The forms shown in Table 3.13 are unmarked for subject person, and by default serve for third-singular subjects. Where subject marking does occur, it is suffixal or enclitic

Table 3.13. *Aspect-mood prefixation Zenzontepec Chatino*

	'darken'	'endure'	'fart'	'have sex (♀)'	'store'	'talk'
POT	ki-kata^{M}	ki^{M}-ku^{H}ti	ki-lya	ki^{M}-kãʔ^{H}	ki^{H}-chã	ki-kwiʔ
PROG	nte-ka^{M}ta^{H}	nte^{M}-ku^{H}ti	nte-lya	ncha^{M}-kãʔ^{H}	nteki^{H}-chã	ncha-kwiʔ
HAB	nti-kata^{M}	nti^{M}-ku^{H}ti	nti-lya	nti^{M}-kãʔ^{H}	nti^{H}-chã	nti-kwiʔ
CPL	nku-ka^{M}ta^{H}	nkwi^{M}-ku^{H}ti	nka-lya	kãʔ^{H}	nkwi^{H}-chã	ya-kwiʔ

	'give'	'lengthen'	'look for'	'rest'	'start'
POT	ta^{M}a^{H}	hni^{H}	tya^{H}na	sukwa^{M}	tyehna^{M}
PROG	nte-ta^{M}a^{H}	nte^{M}-hni^{H}	nte^{M}-tya^{H}na	nte-sukwa^{M}	nte-tyehna^{M}
HAB	n-ta^{M}a^{H}	ntu^{M}-hni^{H}	n-tya^{H}na	nti-sukwa^{M}	n-tyehna^{M}
CPL	nka-ta^{M}a^{H}	nka^{M}-hni^{H}	nkwi^{M}-tya^{H}na	nku-sukwa^{M}	nku-tyehna^{M}

Table 3.14. *Tonal marking of aspect-mood in Zenzontepec Chatino*

	'darken'	'become paralyzed'	'get crushed'	'slip out'
POT	ki-kata^{M}	ki-kela	ki-tita	ki-kili
PROG	nte-ka^{M}ta^{H}	nte-ke^{H}la^{M}	nte-tita	nte-kili
HAB	nti-kata^{M}	nti-kela	nti-tita	nti-kili
CPL	nku-ka^{M}ta^{H}	nku-ke^{H}la^{M}	nku-tita^{M}	nku-kili

Source: Campbell 2011; Campbell, personal communication.

(second person is a tonal marker on the final syllable), and does not interact with the aspect-mood prefixes (Carleton & Waksler 2000).

As will also be apparent, tone may also be an element of tense-aspect inflection. This too falls into different classes which cross-cut the prefixal classes; Table 3.14 shows the tonal variants present in just one of the prefixal classes; the last paradigm represents the type with no tonal inflection, here shown with a toneless verb, but which in principle is possible with any lexical tone. These classes can differ both in the actual tone, and in the shape of the paradigm – the most common patterns being the two-way opposition between potential/habitual and progressive/completive, or the complete absence of a tonal alternation.

For an example of classes based on tense marking, we need look no further than English – namely, the difference between weak and strong past tenses and any further subdivisions within that.

3.6 Conclusion

Inflection classes can be manifested across any feature or combination of features that serves as a basis for inflectional morphology. To the extent that class distinctions result, historically, from purely phonological or morphotactic interactions between elements of the word form, this is hardly a surprising observation. However, where inflection class distinctions (at least, what we choose synchronically to interpret as such) have their origin in semantic or functional categories, we might well expect features to behave differently. It is our impression – intuitively obvious but difficult to quantify – that the more a feature interacts with lexical semantics, the more prone it is cross-linguistically to a variety of morphological realizations as determined by different lexical sets. Number marking is the most prominent in this way, and not just because the morphological marking of number itself is so widespread. Time and again we find languages whose morphology operates on transparent agglutinative principles and yet display a wide variety of number-marking allomorphs. Seri verbs present a particularly striking example. The extreme morphological variety in number marking through suffixation and stem alternations – of which Table 3.7 shows only a portion – contrasts with the consistent prefixal marking of other verbal categories (including subject and object marking, mood, and negation), illustrated in (6).

(6) Ma-ʔ-jo-m-aʔo.
2SG.DO-1SG.SBJ.TR-REAL-NEG-see(SG)
'I didn't see you.' (Marlett 2015: 456)

Although some prefixes display allomorphy morphophonologically conditioned by adjacent segments, there are no lexically specified distinctions.

From the perspective of any individual language, of course, it would be too much to say that we could make any predictions about where to find inflection classes. As stressed in the preceding sections, they have multiple sources, and there is no reason to believe in any kind of systemic oversight that regulates their distribution. But it has obvious ramifications for cross linguistic observations, because the sorts of class distinctions that are in some way related to lexical semantic classification are more likely to be repeated across unrelated languages than those due to language specific phonological or morphological peculiarities.

4 *Motivation*

When we say that lexemes are sorted into different inflection classes, the implicit assumption is that membership in one class or another is arbitrary: the lexemes all share the same morphosyntactic paradigm, and the differences in form have no transparent explanation in terms of any other linguistic property (Corbett 2009a). To take one example from among many, the two classes of verbs in Nez Perce (Table 4.1), a Sahaptin language of the Pacific Northwest (United States.), have different suffixes throughout their paradigm, without there being any synchronically observable semantic or phonological properties that would distinguish the two.

But inflection classes appear to flirt regularly with non-arbitrary assignment principles. This may in part be an echo of their origins, since inflection classes do not arise fully formed, but rather develop from processes of divergence or convergence (see section 2.1). Equally, organizational principles may emerge over time that reduce arbitrariness, as with the Serbo-Croat plural augment *-ov-*, which started off as a characteristic of a lexically specified set of nouns (derived from Indo-European u-stems), but whose use is now substantially conditioned by phonology (see section 5.7 for detailed discussion). As a result, few systems if any maintain complete arbitrariness across the entire relevant set of lexemes. In this chapter, we survey the various supplementary factors that come into play, such as phonology (section 4.1), morphological properties of the paradigm or word form (section 4.2), or semantics (section 4.3). In section 4.4 we look at in-between cases that we identify as *deponent*, where morphology that serves a bona fide morphosyntactic function in one environment appears to be arbitrarily stipulated in another environment, and so is the product of two assignment strategies. Looking within the paradigm, we find that seemingly arbitrary configurations may mask an underlying motivation, so that there is no escape from this ambiguity (section 4.5). In section 4.6 we suggest that maintaining a strict opposition between morphologically arbitrary assignment principles and other types of motivation is still useful, provided one is explicit in tracking their relative contributions to the system.

Table 4.1. *Arbitrarily assigned inflection classes in Nez Perce*

	s-class	c-class
	'eat'	'see'
1SG PRS	hipí-se	hekí-ce
1SG PRF	híp-s	hə́ki-n
1SG PST	híp-e	hə́x-ne
FUT	hip-úʔ	hex-núʔ

Source: Aoki 1994.

Table 4.2. *Phonologically assigned inflection classes in Sobei*

	simple (CV-stem) 'go'	complex (CCV- stem) 'hang'
IRR 1SG	ya-wo	i-mnau
IRR 2/3SG	a-wo	e-mnau
IRR 1INCL PL	ta-wo	te-mnau
IRR 1/2PL	ʔaʔ-wo	ʔeʔ-mnau
IRR 3PL	ria-wo	rie-mnau
REAL 1SG	yo-wo	i-mnau
REAL 2SG	u-wo	u-mnau
REAL 3SG	e-wo	e-mnau
REAL 1INCL PL	ti-wo	te-mnau
REAL 1/2PL	mi-wo	me-mnau
REAL 3PL	ri-wo	re-mnau
SG IMP	a-wo	a-mnau
PL IMP	aʔ-wo	eʔ-mnau

Source: Sterner 1987.

4.1 Phonology

Even if we exclude phonologically *conditioned* variation in morphological form from our picture of inflection class systems, we still find phonological correlations that fall outside the bounds of morphophonology. Consider the paradigm in Table 4.2 from Sobei, a Western Oceanic (Austronesian) language of New Guinea. Subject prefixes fall into two classes, which more or less evenly split the verbal lexicon (43%:57%). The difference is largely predictable from root phonology: the prefix set characteristic of 'go' is used with monosyllabic C-initial verbs, whereas the other set is found with polysyllabic or CC-initial verbs. There is no discernible relationship between the phonological properties of the stems and those of the prefix allomorphs they select: how could the syllable structure of the root affect an alternation in vowel quality? And indeed, as suggested by Sterner (1987), the probable origin of the distinction is not phonological but, rather, morphological. The 'go' type represents monomorphemic verb roots, the 'hang' type represents polymorphemic verb roots with traces of a derivational prefix. On this reconstruction, derived verbs would have taken a different set of mood prefixes (the vowel immediately preceding the root) from underived verbs. As a result, the two classes, though arbitrary in one sense, are nevertheless phonologically predictable.

4.2 Morphology

One obvious way that morphology can influence inflection class membership is by defining properties of the stem that determine inflection class membership. Consider aspectual derivation in Russian verbs illustrated in Table 4.3. There are two main processes. One derives perfective verbs from imperfective through prefixation. The prefixes often add some degree of semantic content, but do not affect membership in the two inflection classes, which is based on the behaviour of the suffixes, as well as associated stem alternations. These prefixed perfective verbs can in turn serve as the base for imperfective verbs derived through a suffix *-yva-/-iva-* (the alternation in the initial vowel is a by-product of the quality of the preceding consonant). Verbs derived with this suffix always belong to the first conjugation. In effect, inflection class membership here is a property of the suffix itself.

Aspects of paradigmatic behaviour can also determine inflection class membership. Consider the noun paradigms in Table 4.4, from Nuer, a Western Nilotic language of South Sudan. Three patterns of suffixation are found in the plural: no suffixation, the suffix *-ni* in the oblique cases, and the suffix *-ni* for all

Table 4.3. *Aspect and inflection class in Russian*

	1st conjugation		2nd conjugation		1st conjugation	
	IPFV	PFV	IPFV	PFV	IPFV	IPFV
	'write'	'rewrite'	'cook'	'overcook'	'rewrite'	'overcook'
INF	pisat´	perepisat´	varit´	perevarit´	perepisyvat´	perevarivat´
1SG NPST	pišu	perepišu	varju	perevarju	perepisyvaju	perevarivaju
2SG NPST	pišeš´	perepišeš´	variš´	perevariš'	perepisyvaeš´	perevarivaeš´
3SG NPST	pišet	perepišet	varit	perevarit	perepisyvaet	perevarivaet
1PL NPST	pišem	perepišem	varim	perevarim	perepisyvaem	perevarivaem
2PL NPST	pišete	perepišete	varite	perevarite	perepisyvaete	perevarivaete
3PL NPST	pišut	perepišut	varjat	perevarjat	perepisyvajut	perevarivajut

Table 4.4. *Plural suffixation in Nuer*

	'fish'		'path'		'orphan'	
	SG	PL	SG	PL	SG	PL
NOM	ré̤ey	rὲc	dúɔɔɔp	dúɔf	rɛ̂ɛt	rɛ̀at-ní
GEN	rɛ́aç	rɛ̂ɛy	dúɔf	dúɔf-nì	rɛ̂at	rɛ̀at-ní
LOC	rɛ́aç	rɛ̂ɛy	dúf	dúɔf-nì	rɛ̂at	rɛ̀at-ní

Source: Irina Monich, personal communication.

cases. These different suffix patterns are related to the pattern of stem alternation. The first two patterns are only possible if there is a singular ~ plural stem alternation (the second pattern is by far the more frequent), whereas the third pattern is obligatory if the stem lacks a number-based alternation (Baerman 2012). The relationship is not entirely deterministic; thus, the third pattern is still possible with nouns that undergo a number-based stem alternation, though it is infrequent. Thus, the suffix classes of the Nuer noun are largely though not completely predictable given the paradigmatic behaviour of stems.

4.3 Semantics

Semantic conditioning presents two interpretative challenges. The first is definitional: since semantic categories may cross the threshold into being features, if a given semantic category reliably correlates with a given inflectional pattern, should it not be incorporated into the underlying feature profile of the paradigm? In that case, we have in effect distinct word classes and not distinct inflection classes. The second challenge comes from the nature of the evidence: sometimes the only overt indication of a semantic contrast is the very morphological contrast we are investigating, so the supposition of semantic predictability can neither be proven nor disproven. We survey here some representative cases.

Morphological variation in nouns in particular often appears to reflect some kind of semantic categorization. For example, nouns in Mali, an East New Britain language of New Guinea, fall into several different classes on the basis of their number marking suffixes. Stebbins (2011) illustrates the semantic contrasts by comparing the same root with different series of number markers. Thus in Table 4.5, the root *amēng* is shared by various words having some connection with wood, differing with respect to such properties as shape and size. (There are two gaps in its distribution in the table: the 'teacher' type is reserved for male humans, whereas the failure of *amēng* to appear in the 'feather' pattern

Table 4.5. *Inflection classes of nouns in Mali*

	'teacher'	'slender tree'	'large full grown tree'	'stick'	'stump'
SG	sunasvem-ka	amēng-ka	amēng-ki	amēng-ini	amēng-vēm
DU	sunasvem-iom	amēng-iom	amēng-vem	amēng-ithom	amēng-vam
PL	sunasvem-ta	amēng	amēng	amēng-ithong	amēng-vap

	'plank'	'pole'	'large log'	'feather'
SG	amēng-igl	amēng-vet	amēng-ia	chēsent-vēs
DU	amēng-iglem	amēng-isem	amēng-inēm	chēsent-imelem
PL	amēng-igleng	amēng-iseng	amēng-inēk	chēsent-imelēk

Source: Stebbins 2011.

is presumably a lexical accident.) The system clearly has some productivity as a derivational device, particularly apparent when we look at borrowings – for example, *plastika* 'medium-sized plastic bottle or bag' ~ *plastiki* 'large-sized plastic bottle or bag' ~ *plastikini* 'small-sized plastic bottle or bag'. Still, it is also lexicalized to some extent, subject at least at times to what in synchronic terms can only be arbitrary principles. Thus, the size contrast implicit in *plastika* vs. *plastiki* is not enough to account for such pairs as *galēp-ka* 'type of fish' ~ *galēp-ki* 'type of palm tree', or *chael-ka* 'wallaby' ~ *chael-ki* 'stingray' (Stebbins & Tayul 2012).

In the Mali system, the categories go hand-in-hand with particular morphological exponents – that is, on the whole, each number marker is at the same time representative of its class.[1] Contrast this with the noun paradigms in Table 4.6 from Diyari, a Pama Nyungan language of Australia. There is a sharp division of inflectional pattern according to semantic principles: common nouns versus women's personal names versus men's personal names. But the contrasting noun classes cannot be encapsulated in contrasting inflectional exponents in the same way as in Mali. For example, both male and female personal names share a suffix *-nha*, but differ in the way it is distributed in the paradigm (absolutive/accusative with male names, accusative, and the base for locative, allative, and dative with female names). Further, the non-singular paradigm of common nouns (here represented by a plural, but the dual is the

[1] Exceptions are nouns ending in *-ka* (belonging to the masculine gender) and *-ki* (belonging to the feminine gender). Nouns in *-ka* follow two patterns, according to whether or not the referent is human: male humans take a plural in *-ta*, while the remaining nouns in *-ka*, and all nouns in *-ki*, share zero marking in the plural.

Table 4.6. *Inflection classes of nouns in Diyari*

	'youth'	'youths'	(woman's name)	(man's name)
ERG	thaṟi-yali	thaṟi-waṟa-li	tyirimiṟi-li	wartamangka-li
ABS	thaṟi	thaṟi-waṟa	tyirimiṟi	wartamangka-nha
ACC	thaṟi	thaṟi-waṟa-nha	tyirimiṟi-nha	wartamangka-nha
LOC	thaṟi-nhi	thaṟi-waṟa-ngu	tyirimiṟi-nhangu	wartamangka-ngu
ALL	thaṟi-ya	thaṟi-waṟa-ngu	tyirimiṟi-nhangu	wartamangka-ngu
DAT	thaṟi-ya	thaṟi-waṟa-rni	tyiṟimiri-nhangka	wartamangka-rni
ABL	thaṟi-ndru	thaṟi-waṟa-ngundru	tyirimiṟi-ngundru	wartamangka-ngundru

Note: [Ch] = dental C, [rC] = retroflex C, [ṟ] = glide /r/ (as opposed to flap or retroflex).
Source: Austin 1981/2013.

same, just with a different stem) patterns with personal names: the oblique case suffixes are those of male names, whereas the behaviour of the core cases (contrast of absolutive and accusative) is that of female names. This means there is no inherent association between the morphological exponents and the lexical categories they are associated with. Instead, the mapping is between the lexical categories and the entire paradigm, which means that the notion of inflection class – as a morphological object – is still useful here, however semantically transparent the assignment rules are.

Cross-linguistically, the question of semantically based noun categorization arises most frequently in connection with gender systems. In many languages the morphological marking of gender agreement overlaps with that of noun morphology. And if gender assignment is at least in part semantically motivated – which it is in any gender system (Corbett 1991: 7–8) – it is natural to try to equate any inflection classes displayed by nouns with the categories described by gender values that have the same morphological expression. In other words, to construe inflection classes as direct exponents of the gender feature. We would argue that this conflates two notions that should be kept distinct, but at the same time shows that otherwise arbitrary inflection class systems can harbour a sizeable element of semantic motivation.

Consider the nouns in Table 4.7 from Eegimaa, an Atlantic language (Niger-Congo) of Senegal. They fall into several inflection classes based on their number prefixes. These in turn map in a non-trivial way onto gender values, which are realized through agreement morphology on verbs and various modifiers (Table 4.8). Sagna (2012) argues that genders I-VI have a definable semantic basis – that is, gender I is for humans, gender II embraces roundness, thickness, and associated notions, and so on. Gender VII is a default category, devoid of

Table 4.7 *Eegima singular~plural noun patterns*

	'builder'	'head'	'kind of animal'	'tree'	'palm tree'	'plank'	'sparrow'
SG	a-ttepa	fu-how	fá-gur	bu-nunuh	ñí-it	ga-babar	ji-ttaja
PL	u-ttepa	gu-how	gá-gur	u-nunuh	ú-it	u-babar	mu-ttaja

	'panther'	'school'	'dog'	'house'
SG	ji-ggaj	Ø-lekkol	e-joba	y-aŋ
PL	si-ggaj	si-lekkol	su-joba	s-aŋ

Source: Sagna 2012.

Table 4.8. *Inflection class – gender mapping in Eegima*

<table>
<tr><th colspan="2">nouns</th><th colspan="3">agreement</th></tr>
<tr><th>SG</th><th>PL</th><th>gender</th><th>SG</th><th>PL</th></tr>
<tr><td>a-</td><td>u-</td><td>I</td><td>na-</td><td rowspan="3">gu-</td></tr>
<tr><td>fu-</td><td>gu-</td><td rowspan="2">II</td><td rowspan="2">fu-</td></tr>
<tr><td>fa-</td><td>ga-</td></tr>
<tr><td>bu-</td><td rowspan="3">u-</td><td>III</td><td>bu-</td><td rowspan="3">u-</td></tr>
<tr><td>ñi-</td><td>IV</td><td>ñu-</td></tr>
<tr><td>ga-</td><td>V</td><td>gu-</td></tr>
<tr><td>ji-</td><td>mu-</td><td>VI</td><td>ju-</td><td>mu-</td></tr>
<tr><td>ji-</td><td rowspan="2">si-</td><td rowspan="4">VII</td><td rowspan="4">e-</td><td rowspan="4">su-</td></tr>
<tr><td>Ø</td></tr>
<tr><td>e-</td><td>su-</td></tr>
<tr><td>y-</td><td>s-</td></tr>
</table>

specific semantic content. If we take the semantic coherence of the first six genders as a given, then in those cases where there is a one-to-one correspondence between the noun's inflection class and gender (gender I, and III-VI), the choice of inflection class for a given noun stem could be said to have a semantic basis. But there are still more inflection classes than there are gender values; thus, the inflection-class distinctions within genders II and VII are underdetermined by the semantic principles observable in the gender system, and hence they are arbitrary.

Table 4.9. *Fragment of the Rotokas subject-marking paradigm, habitual present forms*

	alpha class	beta class
	'go'	'eat'
1	avapa-ra-i	aiopa-a-voi
2	avapa-u-ei	aiopa-ri-voi
3M	avapa-ro-i	aiopa-re-voi

Note: First and second person forms of 'eat' have been partially constructed.
Source: Robinson 2011.

The question of semantic categorization as a component of inflection class also arises with verbs, but because of the nature of verbal paradigms – or at least, our conventional interpretations of them – the issues are somewhat different. The semantic categories involved in inflectional contrasts in verbal paradigms are often ones that implicate the morphosyntactic paradigm, such as argument structure and role marking. The result is categories that line up approximately with syntactic distinctions, but not seamlessly; so it is not clear, at least under a modular conception of language, which component of grammar is responsible.

As an example, consider the inflectional contrast that pervades the verbal system of Rotokas, a language of the North Bougainville family of New Guinea. Table 4.9 shows a representative fragment of the larger paradigm, involving dual and plural subject number and further TAM distinctions. (The first suffix here marks subject person-number, the second marks TAM.) Following Robinson (2011), we call these the alpha and beta class.

Examples (7)–(10) illustrate the contrast in use between the two inflectional types. In (7) a transitive verb with a direct object inflects according to the beta-class. When the same verb is detransitivized (taking an incorporated object, so that the notional object is now marked as an oblique, with the locative case), it inflects according to the alpha class (8). The intransitive verb 'go' in (9) also inflects according to the alpha class. So far, this suggests we interpret the apha-class suffixes as markers of intransitivity, and the beta-class suffixes as markers of transitivity. But we also get verbs like 'walk' in (10), which are intransitive but inflect according to the 'transitive' class.

(7)	Rarasori	kakapiko-a	aio-a	aio-pa-re-voi
	Robinson	little-SG.N	food-SG.N	eat-CONT-3SG.M-PRS

'Robinson eats little food.' (Robinson 2011: 201)

(8)

Reari	ira	akoroa=ia	aasi	aio-pa-ro-i
Reari	RPRO.3.SG.M	betel net=LOC	betel.nut	eat-CONT-3SG.M-PRS

'Reari is chewing on betel nut.' (Robinson 2011: 201)

(9)

viapau	ava-pa-ro-i	eisi	kovo-a
NEG	go-CONT-3SG.M-PRS	LOC	garden-SG.N

'…he doesn't go to the garden.' (Robinson 2011: 92)

(10)

Jon	kovuru-vira	voka-pa-re-voi	raiva=ia
John	cross-ADV	walk-CONT-3SG.M-PRS	road=LOC

'John walks across the road.' (Robinson 2011: 210)

Rotokas is thus a classic example of 'split intransitivity'. Transitive verbs display uniform behaviour, but intransitives are split: some behave like transitives, others behave inflectionally according to a pattern distinct from transitives. Robinson (2011), along with many others, speculates that there is a semantic basis for this split, along a cline of semantic transitivity that is separate from actual syntactic valence. But the actual semantic factors remain elusive, which means that at least provisionally it may be better to treat this as an inflection class contrast that has a predominantly syntactic and/or semantic application, rather than as an imperfectly characterizeable component of the morphosyntactic paradigm. In the following section, we explore these borderline phenomena in more detail.

4.4 Deponency

A modular conception of language is one of the noteworthy achievements of the discipline of linguistics. Being able to divide an analysis into phonological, morphological, syntactic, lexical, and so forth elements lends clarity and cross-linguistic consistency to a description. But, in practice, this model is often an idealization over a body of facts that do not quite line up, something we should bear in mind when considering what exactly it is that lies behind some morphological form.

As an illustration of the issues involved, consider the following examples of case-number inflection from the Baltic (Indo-European) language Latvian. Table 4.10 shows some nouns that are divided into four inflection classes based on their suffixation. All the nouns are inanimate, and assignment to these classes is arbitrary (as far as anyone can tell).

The same inflection classes are found with animate nouns, but in this case there is a clear relationship to the sex of the referent: inflection classes 1

Table 4.10. *Inflection class distinctions in inanimate nouns in Latvian*

	1st decl.	4th decl.	2nd decl.	5th decl.
	'hockey'	'aerobics'	'shelf'	'ticket'
NOM SG	hokej-s	aerobik-a	skap-is	bilet-e
ACC SG	hokej-u	aerobik-u	skap-i	bilet-i
GEN SG	hokej-a	aerobik-as	skapj-a	bilet-es
DAT SG	hokej-am	aerobik-ai	skap-im	bilet-ei
LOC SG	hokej-ā	aerobik-ā	skap-ī	bilet-ē
NOM PL	hokej-i	aerobik-as	skapj-i	bilet-es
ACC PL	hokej-us	aerobik-as	skapj-us	bilet-es
GEN PL	hokej-u	aerobik-u	skapj-u	bileš-u
DAT PL	hokej-iem	aerobik-ām	skapj-iem	bilet-ēm
LOC PL	hokej-os	aerobik-ās	skapj-os	bilet-ēs

Source: Mathiassen 1997.

Table 4.11. *Inflection class and sex distinctions in human referent nouns in Latvian*

	1st decl.	4th decl.	2nd decl.	5th decl.
	'reader (M)'	'reader (F)'	'Latvian (M)'	'Latvian (F)'
NOM SG	lasītāj-s	lasītāj-a	latviet-is	latviet-e
ACC SG	lasītāj-u	lasītāj-u	latviet-i	latviet-i
GEN SG	lasītāj-a	lasītāj-as	latvieš-a	latviet-es
DAT SG	lasītāj-am	lasītāj-ai	latviet-im	latviet-ei
LOC SG	lasītāj-ā	lasītāj-ā	latviet-ī	latviet-ē
NOM PL	lasītāj-i	lasītāj-as	latvieš-i	latviet-es
ACC PL	lasītāj-us	lasītāj-as	latvieš-us	latviet-es
GEN PL	lasītāj-u	lasītāj-u	latvieš-u	latvieš-u
DAT PL	lasītāj-iem	lasītāj-ām	latvieš-iem	latviet-ēm
LOC PL	lasītāj-os	lasītāj-ās	latvieš-os	latviet-ēs

Source: Mathiassen 1997.

and 2 are masculine, and classes 4 and 5 are feminine (Table 4.11). Does that mean we should treat these inflection classes as markers of sex, at least where the referent can be so characterized? Perhaps, however, the relationship is indirect. Alongside pairs as in Table 4.11, where we can think in terms of a single stem that is inflected according to one or the other inflectional pattern, we have pairs such as *amerikāniet-is* (M), *amerikānet-e* (F) 'American', which display distinct stems. This suggests that underlying the

Table 4.12. *Short form (indefinite) adjective 'big' in Latvian*

	masculine	feminine
NOM SG	liel-s	liel-a
ACC SG	liel-u	liel-u
GEN SG	liel-a	liel-as
DAT SG	liel-im	liel-ai
LOC SG	liel-ā	liel-ā
NOM PL	liel-i	liel-as
ACC PL	liel-us	liel-as
GEN PL	liel-u	liel-u
DAT PL	liel-iem	liel-ām
LOC PL	liel-os	liel-ās

pairs in Table 4.11 is a null derivational process that toggles between masculine and feminine stems.[2]

These inflection classes in turn have a non-trivial relationship to adjectival inflection: the opposition of 1st ~ 4th declension is what marks the distinction between masculine and feminine in short form (indefinite) adjectives (Table 4.12).[3] In these paradigms we are forced to say that this morphological opposition realizes the morphosyntactic feature of gender.

So in the end we have a similar set of morphological oppositions operating in three domains: as arbitrary inflection class attributes with non-human nouns, as sex markers with human nouns, and as gender markers with adjectives. Which function is primary, or does it even make any sense to pose a question like that? In practice there is no reigning consensus about how to deal with this question, although it might be reasonable to assume that the primary function would be associated with the word class that is cross-linguistically basic (i.e., nouns in this case).

Probably the most familiar example of this kind of analytical ambiguity are deponent verbs in Latin. Transitive verbs regularly oppose active and passive forms, as the partial paradigms in Table 4.13 shows. Deponent verbs – typically intransitive, though there are a few transitives as well, as shown here – use passive forms (where available in the paradigm), but are not otherwise

[2] Also, note that there are unpaired roots with a masculine referent that inflect like feminine nouns, except for their dative singular; thus *poika* 'boy' inflects as class 5, except for dative singular *poikam*, which takes the class 1 suffix.

[3] Long form (definite) adjectives have different suffixes, going back historically to the suffixation of a demonstrative modifier.

Table 4.13. *Partial paradigm of 'normal' and deponent transitive verbs in Latin*

	active	*passive*	*deponent*
	'love'	'be loved'	'exhort'
PRS IND 1SG	amō	amor	hortor
PRS IND 2SG	amās	amāris	hortāris
PRS IND 3SG	amat	amātur	hortātur
PRS IND 1PL	amāmus	amāmur	hortāmur
PRS IND 2PL	amātis	amāminī	hortāminī
PRS IND 3PL	amant	amantur	hortantur
IMP SG	amā	amāre	hortāre
INF	amāre	amārī	hortārī
PRF IND 1SG	amāvī	amātus sum	hortātus sum

Table 4.14. *The major inflection classes of Latin verbs, active forms*

	'love'	'advise'	'rule'	'hear'
PRS IND 3SG	amat	monet	regit	audit
IMPRF IND 3SG	amābat	monēbat	regēbat	audiēbat
FUT IND 3SG	amābit	monēbit	reget	audiet
PRS SBJV 3SG	amet	moneat	regat	audiat
IMPRF SBJV 3SG	amāret	monēret	regeret	audīret

demonstrably 'passive'. Whether a verb is deponent or not appears to be lexically specified (though many share semantic properties that suggest they might be treated as 'middle voice'; Kemmer 1993). On the face of it, there are two equally satisfying (or unsatisfying) analytic alternatives: (1) deponent verbs are lexically specified as inflecting like passives, even though they are not, or (2) active and passive inflection represents two inflection classes; transitives employ the opposition between these two classes to mark the active ~ passive alternation.

The traditional approach is the first one, probably due to consideration of a further morphological fact. Latin also has a distinction between four major inflection classes (Table 4.14). These were originally due to the stem-final vowel rather than any differences in the inflectional suffixes, and some portions of the paradigm still lend themselves to such an analysis – for example, the imperfect subjunctive third-person singular forms in Table 4.14 could easily

Table 4.15. *The major inflection classes of Latin verbs, passive forms*

	'be loved'	'be advised'	'be ruled'	'be heard'
PRS IND 3SG	amātur	monētur	regitur	audītur
IMPRF IND 3SG	amābatur	monēbatur	regēbatur	audiēbātur
FUT IND 3SG	amābitur	monēbitur	regētur	audiētur
PRS SBJV 3SG	amētur	moneātur	regātur	audiātur
IMPRF SBJV 3SG	amārētur	monēret	regerētur	audīretur

be segmented as *amā-ret*, *monē-ret*, *rege-ret*, and *audī-ret*, with four different stem-final vowels and an invariant suffix, just as in Wintu (section 2.2). But elsewhere in the paradigm matters become more opaque in synchronic terms. For example, on the basis of the last three classes, the present subjunctive 3SG appears to involve shortening of a long or deletion of a short stem-final vowel (**monē+at* → *moneat*, **audī-at* → *audiat*, **rege+at* → *regat*), but 'love' would require an ad hoc morphophonological rule transforming a sequence of /a+a/ into /e/. Worse, the future indicative forms of the 'love' and 'advise' classes take an additional stem augment *-b-* which is lacking in the 'rule' and 'hear' classes. Thus, the system has the distinctions in inflectional morphology characteristic of inflection classes.

These distinctions are faithfully mirrored in the passive forms (Table 4.15). That means if we treat active and passive as parallel but separate sets of lexical items in some kind of derivational relationship, the coordination of their inflection classes emerges as a freakish accident. Alternatively, if we allow inflection class membership to be transferred across derivationally related items, we have in effect said that they are *not* derivationally related in any meaningful sense. This interdependence suggests that we treat active and passive forms as constituting part of a single paradigm, with deponent verbs representing a particular deviation from this (see section 5.5.8 for further discussion).

This is not to say that all superficially similar examples lend themselves to the same interpretation. A case in point is subject suffixation on verbs in Lealao Chinantec, an Oto-Manguean language of Mexico. In this language, animacy is an agreement feature in some parts of the grammar (e.g., determiners, numerals), but with verbs its grammatical status is more uncertain. Most verb roots are restricted to having either an animate or an inanimate 'absolutive' argument, i.e., intransitive subject or transitive object. But some roots occur in animacy pairs – that is, their lexical semantics are such as to be compatible with both. Table 4.16 illustrates unpaired and paired verbs, both intransitive

Table 4.16. *Animate versus inanimate verbs in Lealao Chinantec*

a. intransitive

	unpaired		paired	
	animate	*inanimate*	*animate*	*inanimate*
	'dance'	'blossom'	'(someone) falls over'	'(something) falls over'
1SG	ʔi^{2}bi^{32}dxá^{1}-á^{4}	——	ʔi^{4}tiáʔ^{4}-á^{4}	——
2SG	ʔi^{2}bi^{32}dxá^{1}-u^{3}	——	ʔi^{4}tiáʔ^{4}-u^{3}	——
3SG	ʔi^{2}bi^{32}dxá^{1}	ʔí^{4}lí^{1}	ʔi^{4}tiá<y>ʔ^{4}	ʔi^{4}tiaʔ^{3}

b. transitive

	unpaired		paired	
	animate	*inanimate*	*animate*	*inanimate*
	'bother'	'weave'	'push sb.'	'push sth.'
1SG	ʔi^{2}bi^{3}dxɨɨ^{1}-á^{4}	ʔi^{2}ta^{2}-y	ʔi^{2}liaʔ^{1}-á^{4}	ʔi^{2}lia<y>ʔ^{42}
2SG	ʔi^{2}bi^{3}dxɨɨ^{1}-u^{3}	ʔi^{2}ta^{2}-y	ʔi^{2}liaʔ^{4}-u^{3}	ʔi^{2}lia<y>ʔ^{3}
3SG	ʔi^{2}bi^{3}dxɨɨ^{1}	ʔi^{2}ta^{2}	ʔi^{4}liá<y>ʔ^{2}	ʔi^{4}liáʔ^{4}

Note: Suffixal -*y* is realized prior to any stem-final /ʔ/.
Source: Analysis from Palancar 2014, data from Rupp & Rupp 1996.

Table 4.17. *The three major inflection classes of Lealao Chinantec verbs (96% of the fully inflected verbal lexicon)*

	class I	class II	class III
	'tie something'	'receive something.'	'run someone over'
1SG	ʔi^{2}ñuu^{42}-y	ʔi^{4}tǿn^{4}-á^{4}	ʔi^{2}heé^{2}-á^{4}
2SG	ʔi^{2}ñuu^{3}-y	ʔi^{4}tǿn^{4}-u^{3}	ʔi^{2}heé^{2}-u^{3}
3SG	ʔi^{4}ñuú^{4}	ʔi^{4}tǿn^{4}	ʔi^{4}heé^{2}-y

and transitive. The contrasting suffixation will be the topic of the following discussion. (Note that there are also plural forms, but these are transparently constructed through affixation to the forms already illustrated here, and so would add nothing to the discussion; the forms shown are future/potential, which is the citation form in Rupp & Rupp's 1996 dictionary.)

Against this background, consider now the three major inflection classes (Table 4.17), which account for 96 percent of the fully inflected verbal lexicon (i.e., verbs that allow inflection for all three subject persons). There is a

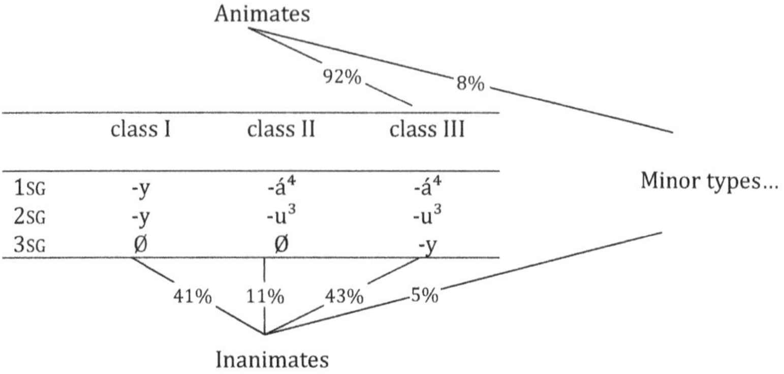

Figure 4.1. *Mapping of animacy type onto inflection class in Lealao Chinantec: unpaired transitives*

non-trivial relationship between animacy and inflection class, but it is a complex one. For the sake of simplicity, we look here only at transitive verbs, since intransitive verbs obviously lack complete paradigms in the case of inanimate subjects.

With unpaired verbs, animate verbs usually inflect as class III, whereas inanimate verbs show no particular tendency, as shown in Figure 4.1. With paired verbs, animates *always* inflect as class III, whereas inanimates are overwhelmingly assigned to classes I or II, as shown in Figure 4.2. Within this particular domain, the assignment of inanimates to class I or II is itself largely predictable by morphological criteria – namely, the behaviour of tone: if both members of the pair have the same tone, the inanimate verb usually inflects as class II. Other verbs undergo a tonal alternation. In that case the inanimate verb usually inflects as class I (Table 4.18).

Thus within a narrowly defined domain we can say that this morphological opposition realizes the morphosyntactic feature of animacy, but with respect to the verb system as a whole it is perhaps better classed as simply an inflection class opposition. This is not however to suggest that the Latin and Lealao Chinantec examples are *fundamentally* different; rather it is a necessary evil engendered by our (in general justified) analytic conventions that we are forced to anchor our description somewhere, either in the inflectional categories, or in the morphological paradigms. But we should recognize the fluidity of the boundaries.

In the foregoing examples, evidence for dual patterning comes from a comparison across lexemes – for example, the syntactic and semantic attributes of

Table 4.18. *Inflection class selection correlated with tone in Lealao Chinantec*

	pair without tonal alternation		pair with tonal alternation	
	animate (III)	inanimate (II)	animate (III)	inanimate (I)
	'receive someone'	'receive something'	'push someone'	'push something'
1SG	ʔi^{4}tǿn^4-á^4	ʔi^{4}tǿn^4-á^4	ʔi^{2}liaʔ^1-á^4	ʔi^{2}lia<y>ʔ^{42}
2SG	ʔi^{4}tǿn^4-u^3	ʔi^{4}tǿn^4-u^3	ʔi^{2}liaʔ^4-u^3	ʔi^{2}lia<y>ʔ^3
3SG	ʔi^{4}tǿn-y^4	ʔi^{4}tǿn^4	ʔi^{4}liá<y>ʔ^2	ʔi^{4}liáʔ^4

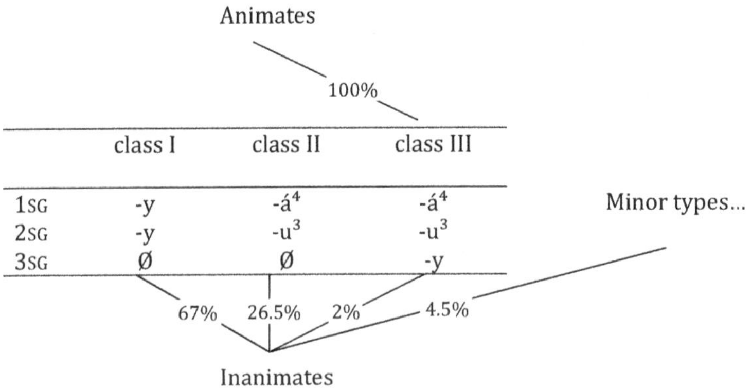

Figure 4.2. *Mapping of animacy type onto inflection class in Lealao Chinantec: paired transitives*

'passive' morphology with the verb 'love' are different from the verb 'exhort'. In some cases the evidence is actually paradigm-internal. Consider the system of subject prefixes in Santa Ana Keres, a Keresan language of New Mexico (Table 4.19). These follow either of two patterns, one characterized as actor marking, the other as undergoer marking, which is allied with object marking. It is not entirely clear how deterministic the semantic characterization of these two patterns is or whether there is some degree of lexical specification involved. But whatever the case may be, a verb like 'eat' in Table 4.19 stands out: it behaves like an 'actor' verb in the singular and like an 'undergoer' verb in the plural. Whatever semantic principles might underlie the distinction, they should not change along with subject number.

Table 4.19. *Santa Ana Keres verbal prefixes*

	actor 'work'	undergoer 'fast'	?? 'eat'
1SG	s-utaniça	s-udʸaši	s-upe
2SG	ṣ-utaniça	gəẓ-udʸaši	ṣ-upe
3SG	k-utaniça	ga-udʸaši	k-upe
1PL	s-uwataniẓanˀe	səẓ-uwadʸaši	səẓ-a:ʔape
2PL	ṣ-uwˀataniẓanˀe	gəẓ-uwadʸaši	gəẓ-aʔape
3PL	g-uwˀataniẓanˀe	g-uwadʸaši	g-aʔape

Source: Davis 1964.

In Takelma, a language formerly spoken in Oregon (classified by some as a member of the Penutian family), values of person provide a context for a similar morphological alternation. The major inflectional patterns are shown in Table 4.20. Intransitive verbs fall into two main inflection classes based on their subject-marking suffixes, characterized by Sapir (1922) as active (class I) and stative (class II). Transitive verbs (class III) have a distinct subject-marking paradigm, whose suffixes are either different from either class I or II, or correspond to the active suffixes. (First and second person objects are also overtly marked, by additional suffixes.) In addition, there is a fourth class. These are intransitive verbs that follow the stative pattern in the second and third person, but in the first person they additionally take the causative suffix *-na-*, and then inflect as transitives. If we attribute meaning to the first three classes (active, stative, transitive), the pattern in class IV makes little sense – particularly as it involves an alternation between the least active and most active types, skipping the intermediate class.

Such paradigm-internal irregularities are relatively unsurprising if we think of these paradigms as representing inflection classes: they are then instances of heteroclisis (Stump 2006), which are hardly problematic from an analytical point of view. But given the apparently close alignment of the morphological paradigms with semantically and syntactically identifiable categories elsewhere in the system, we should not be too quick to dismiss the oddness of the situation.

4.5 Paradigm Shape and Morphosyntactic Function

The fluid boundary between morphological arbitrariness and morphosyntactic motivatedness that deponency represents has an analogue in the way

Table 4.20. *Inflection classes of verbs in Takelma*

	I active INTR 'come'	II stative INTR 'stop'	III TR (3rd person object) 'kill'	IV?? 'work'
AOR 1SG	baxam̌-t^{h}eʔ	hanâʔs-teʔ	tʔomomâ-ʔn	hekwêhak^{hw}-na-ʔn
AOR 1PL	baxam-ík^{h}	hanâʔs-ik^{h}	tʔomoma-nâk^{h}	hekwêhak^{hw}-na-nák^{h}
AOR 2SG	baxamá-t^{h}	hanâʔs-tam	tʔomomá-t^{h}	hekwêhak^{hw}-tam
AOR 2PL	baxamá-t^{h}p^{h}	hanâʔs-tap^{h}	tʔomomá-t^{h}p^{h}	hekwêhak^{hw}-tap^{h}
AOR 3	baxâʔm	hanâʔs	tʔomóm	hekwêhak^{hw}
FUT 1SG	baxmâ-t^{h}e:	hânʔs-te:	to:mâ-n	he:kwâk^{hw}-na-n
FUT 1PL	baxma-kâm	hânʔs-ikam	to:ma-nakâm	he:kwâk^{hw}-na-nakam
FUT 2SG	baxma-tâʔ	hânʔs-taʔ	to:ma-tâʔ	he:kwâk^{hw}-taʔ
FUT 2PL	baxmâ-t^{h}paʔ	hânʔs-tapaʔ	to:mâ-t^{h}paʔ	he:kwâk^{hw}-tapaʔ
FUT 3	baxmâ-ʔt^{h}	hânʔs-ta:	to:má-nk^{h}	?[a]
POT 1SG	baxmâ-t^{h}eʔ	hânʔs-teʔ	tó:mâ-ʔn	he:kwâk^{hw}-na-ʔn
POT 1PL	baxm-ík^{h}	hânʔs-ik^{h}	tó:ma-nâk^{h}	he:kwâk^{hw}-na-nák^{h}
POT 2SG	baxmâ-t^{h}	hânʔs-tam	tó:má-t^{h}	he:kwâk^{hw}-tam
POT 2PL	baxmá-t^{h}p^{h}	hânʔs-tap^{h}	tó:má-t^{h}p^{h}	he:kwâk^{hw}-tap^{h}
POT 3	baxmâ-ʔ	hânʔs	tó:m	he:kwâk^{hw}

[a] Sapir (1922) does not provide information about this cell.
Source: Sapir 1922.

individual paradigms are configured. Consider first the paradigm in Table 4.21, from Chipaya, an Uru-Chipaya language of Bolivia. This shows the declarative subject agreement enclitic, which docks onto various hosts. The paradigm distinguishes values of person, number, and gender, but the forms *il* and *izh* can only be described in terms of a disjunction of values: *il* is used for first-person (exclusive) and third-person singular feminine, whereas *izh* is used for third-person masculine and all the plural values other than first-person exclusive.[4] Because disjunctive patterns such as this cannot be described as a natural class of feature values, it looks to be morphologically arbitrary.

But not all disjunctive patterns can be quite so readily ascribed to morphological stipulation. Precisly here, at the intersection of person, gender, and number, we find a whole family of examples where apparent disjunction may be motivated at another level. Consider the verbal paradigm in Table 4.22, from Tucano, a Tucanoan language of Columbia. Verbs mark subject agreement for person, number, and gender; for the third person, masculine, feminine, and neuter are distinguished. First and second person subjects, however, do not distinguish gender, and, surprisingly, the form used for them is the same as the third person neuter – that is, precisely the gender value that would *not* apply

[4] 1INC PL and 2PL optionally take distinct enclitics, *trum* and *truk*, respectively.

Table 4.21. *Chipaya declarative subject agreement enclitic*

	SG	PL
1INCL	–	=izh
1	=il	=il
2	=im	=izh
3 M	=izh	=izh
3 F	=il	=izh

Source: Cerrón-Palomino 2006: 172–175.

Table 4.22. *Tucano 'do'*

<table>
<tr><th></th><th>SG</th><th>PL</th></tr>
<tr><td>1</td><td colspan="2" rowspan="3">wee-ʔe</td></tr>
<tr><td>2</td></tr>
<tr><td>3 N</td></tr>
<tr><td>3 M</td><td>wee-mí</td><td rowspan="2">wee-má</td></tr>
<tr><td>3 F</td><td>wee-mó</td></tr>
</table>

Source: West & Welch 2004: 37.

to first- and second-person arguments. Thus, this form involves the disjunction of first and second person, by implication non-neuter, with the third-person neuter.

One could, however, understand this as reflecting the intersection of morphological and morphosyntactic asymmetries. The morphological asymmetry is manifested in the treatment of gender marking with the third person: only masculine and feminine are overtly represented, whereas the form used with neuter subjects is simply a default form, compatible in principle with anything, but effectively limited to neuter arguments. The morphosyntactic asymmetry is manifested in the syntactic properties of different persons: third-person arguments control gender and number agreement, but first- and second-person arguments do not. On this interpretation, first- and second-person subjects take the default form, which also happens to be the one used for third person neuter subjects.

On the face of it, this might seem to be a baroque explanation for a minor paradigmatic quirk, but the pattern in its essentials is repeated across various unrelated languages, increasing its plausibility. For example, Krongo, a Kadugli language of Sudan (Table 4.23), has essentially the same pattern as

Table 4.23. *Krongo 'saw'*

	SG	PL
1		
2	n-àasàlà	
3 N		k-àasàlà
3 M	àasàlà	
3 F	m-àasàlà	

Source: Reh 1985: 186.

Tucano, with the difference that plural agreement applies to all persons, and not just third-person masculine and feminine.

Although the assumption that the first-/second-/third-person neuter form is a morphological default requires a certain suspension of disbelief (there is no evidence for it other than the contingent demands of analyzing these particular paradigms), the restriction of gender agreement to third person does mirror a cross-linguistically familiar restriction of pronominal gender distinctions to third person (Siewierska 2004: 104–105). That said, this paradigmatic type cannot necessarily be construed as a translation of pronominal regularities, because further peculiarities may be found. For example, in some Nakh-Daghestanian languages, the person-based restriction on gender agreement occurs only in the plural. In most of the languages of this family there is no person agreement as such, only gender and number agreement (with the absolutive argument). Consider the Archi copula in Table 4.24. In the singular, four genders are manifested: genders I and II for humans, and genders III and IV for non-humans. In the plural, humans are distinguished from non-humans, but only for third-person arguments; first and second person take the same agreement marking as non-humans. A similar pattern is found in Ingush (Nichols 2011: 143, 431); this is striking, because although both languages are from the same family their relationship is remote, Archi being from the Daghestanian branch and Ingush from the Nakh branch, and thus of a degree of separation comparable to the different branches of Indo-European (Nichols 2015: 270).

This interpretation assumes a certain separation between morphological forms and the values they represent: the forms realize certain morphosyntactic values, but the way they are deployed in the paradigm interacts with restrictions imposed by other feature values. The fluid nature of this relationship is strikingly illustrated by Orejón, like Tucano a Tucanoan language, but from

Table 4.24. *Archi 'be.PRS'*

singular		plural	
I (male)	w-i	1	Ø-i
II (female)	d-i	2	
III	b-i	III	
IV	Ø-i	IV	
		I (male 3rd person)	b-i
		II (female 3rd person)	

Source: Corbett 2012: 239–251, based on Kibrik et al. 1977.

Table 4.25. *Orejón present-future suffixes*

	indicative		interrogative	
	F	M	F	M
PL	-yo		-ye	
1SG	-yi		-yi	
2SG			-ko	-kɨ
3SG	-ko	-hɨ		

Source: Velie & Velie 1981: 123f.

the Western rather than the Eastern branch. Before highlighting the relevant points, it should be noted that Orejón differs from the languages presented so far, in that there are only two genders, masculine and feminine, and nouns that denote inanimates take masculine agreement. With that in mind, consider first the indicative present-future paradigm in Table 4.25. This is, in effect, the two-gender analogue of the Tucano paradigm, with gender agreement in the third-person singular, and one form for the rest of the singular. Contrast this with the corresponding interrogative paradigm. Each paradigm comprises four suffixes that, although not identical (two of the four differ slightly), are no doubt morphologically related. But the striking fact is that their distribution is different: whereas in the indicative the gender-agreeing suffixes are restricted to the third-person singular, in the interrogative their range is extended to the second-person singular. This can be seen even more clearly in the past tense paradigms in Table 4.26, which have only three forms each: two gender-agreeing forms, and a single form for the rest. The indicative and interrogative paradigms thus are characterized by different distributions (and partly different forms, it must be admitted) of the gender markers based on which person values are licit agreement controllers in these contructions.

Table 4.26. *Orejón past suffixes*

	indicative		interrogative	
	F	M	F	M
PL 1SG	-bɨ		-de	
2SG 3SG	-go	-gɨ	-go	-gɨ

Source: Velie & Velie 1981: 123f.

Table 4.27. *Tucano nominal forms*

	a. gerundive 'wash'		b. comparable suffixes on nouns	
	SG	PL		
M	coe-gʉ	coe-rã	acaweré-gʉ	'male relative'
F	coe-go		acaweré-go	'female relative'
N	coe-ro		acaweré-rã	'relatives'
			acá-ro	'box'

Source: West & Welch 2004: 37, 81, 85.

On the other hand, we should not get too carried away with an account driven by twists in the rules of agreement. Consider Tucano again. Many verbal constructions involve a nominal form, termed gerundive in the description. The nominal form marks gender-number using suffixes identical to those found on nouns, as in Table 4.27, part (b).[5] This gerundive forms a periphrastic construction together with an auxiliary verb (the verb 'do' shown earlier in Table 4.22). But whereas the auxiliary displays the apparent person-based restrictions on gender agreement, the gerundive does not. The result is a periphrastic construction, such as that shown in Table 4.28 (what we interpret as the default form of the auxiliary is shown underlined), whose individual members display different gender agreement patterns. If we treat this construction as a single agreement target, then clearly the gender restriction is morphological and not morphosyntactic.

[5] The noun system includes a large number of different singular and plural suffixes, but gerundive inflection is limited to this set of four. Note that inanimate count nouns typically have a distinct plural form (e.g., *acá-ri* 'boxes'), but always take singular agreement.

Table 4.28. *Tucano present progressive paradigm (gerundive + auxiliary) 'is washing'*

<table>
<tr><th></th><th>SG</th><th>PL</th></tr>
<tr><td>1 M</td><td rowspan="2">coe-gʉ wee-ʔe</td><td rowspan="6">coe-rã wee-má</td></tr>
<tr><td>2 M</td></tr>
<tr><td>3 M</td><td>coe-gʉ wee-mí</td></tr>
<tr><td>1 F</td><td rowspan="2">coe-go wee-ʔe</td></tr>
<tr><td>2 F</td></tr>
<tr><td>3 F</td><td>coe-go wee-mó</td></tr>
<tr><td>3 N</td><td colspan="2">coe-ro wee-ʔe</td></tr>
</table>

Source: West & Welch 2004: 37.

Table 4.29. *Carapana non-past conjectural 'work'*

<table>
<tr><th></th><th>SG</th><th>PL</th></tr>
<tr><td>1 M</td><td rowspan="2">paa-ʉ</td><td rowspan="6">paa-rã</td></tr>
<tr><td>2 M</td></tr>
<tr><td>3 M</td><td>paa-ʉ-mi</td></tr>
<tr><td>1 F</td><td rowspan="2">paa-o</td></tr>
<tr><td>2 F</td></tr>
<tr><td>3 F</td><td>paa-o-mo</td></tr>
<tr><td>3 N</td><td colspan="2">paa-ro</td></tr>
</table>

Source: Metzger 2000: 154.

In most other Eastern Tucanoan languages the auxiliary element is suffixed to the nominal form; this means that the morphological unity of the construction is even more apparent, as in the non-past conjectural paradigm of Carapana in Table 4.29. Note here that the auxiliary – that is, the equivalent of *wee* in Tucano – is simply zero.

The paradigmatic configurations surveyed in this section show us again that the distinction between morphosyntactically motivated and morphologically arbitrary is not always clear-cut. The disjunctive distribution of forms within these paradigms, which ought to be a sign of arbitrariness, may well have a morphosyntactic motivation, in the sense that morphosyntactic features may license agreement. At the same time though, we should not allow ourselves to be completely seduced by syntactic explanations, as the dual agreement patterns in auxiliary constructions in Tucano and Carapana show. This suggests that, sometimes at least, we face not a choice between competing accounts, but rather the simultaneous validity of both.

4.6 Conclusion

In this work, we understand morphological complexity as arising where inflectional morphology strays out of bounds, those bounds being defined somewhere other than in morphology proper, such as in syntax or semantics. We can treat this as an all or nothing affair: forms either follow the terms defined by syntax and semantics or they do not. On this simplifying assumption, calling something an inflection class means it is wholly arbitrary and must be explicitly stipulated. By the same token, the distribution of forms in a paradigm should be definable in terms of natural classes of morphosyntactic feature values – otherwise it is arbitrary and morphologically stipulated. The facts, as so often is the case, do not always cooperate with this simple dichotomy. There are seeming inflection classes that nevertheless have some relation to meaning or function, and there are paradigms whose distribution of forms is certainly related to morphosyntactic values but where this relationship is indirect. One could take this as evidence that a motivated versus arbitrary distinction is artificial, and we need a more gradient and nuanced view of how morphological paradigms come about. Or one could maintain the distinction, and take these data as a call for a more precise mapping between the two, say by weighting the differing contributions coming from either side of the grammar. From a descriptive and typological point of view, we suggest maintaining the latter approach, because the terms and analytic concepts we currently have at our disposal presuppose it.

5 *Conditions on Paradigms*

When analysing complex data, we often set aside one part of the data if this allows us to understand another. We may, for instance, put stems temporarily to one side in a given system, while we get a grasp of inflectional affixes. And we may then try to hold the affixes constant, while we work out the behaviour of stems. This is a valid strategy, used widely across science. But having gained a clearer understanding of the parts, we then have to see how they integrate. In our case, we want to know how the system of affixes is integrated with the system of stems. They may match up, they may cross-cut, or else we may find that they interrelate in a more complex way; in other words, they are not independent of each other and yet neither fully predicts the other. Of course, we have to guard against setting aside some relevant parts of the data and then failing to return to them. While trying to understand the core elements, morphologists often abstract away from material that appears to be outside the system. Alternations may be (partly) predictable from external factors. Yet, as we shall see, this material too may again be systematic, or less systematic. We find new generalizations here, and then go on to integrate them with our understanding of the core.

We can see the analyses in the previous chapters as being different ways of partitioning the lexicon into classes of items: those items with a particular affix versus the others, those with a stem alternation versus the others, and so on. But there are further ways in which inflection is determined, different sources of criteria for partitioning the lexicon. Some of these are further away from core morphology. It is time to tackle these systematically.

The generalizations we shall examine are termed 'conditions on inflection'. Such generalizations are often relegated to the footnotes in the description of inflectional systems. And in our discussion so far, they have been set aside deliberately, rather than being at the centre of attention. We shall see that it makes sense to treat these generalizations separately, since there are examples where they cross-cut the more familiar generalizations of inflection classes. They also turn out to be rather interesting in their own right. Indeed they form the basis for a typology, which proves surprisingly complete.

Their importance for complexity has already been hinted at. Since they cross-cut the types of generalization which we have established as necessary for describing inflection systems in the preceding chapters, they allow us to approach the complexity of systems by looking at the different types of information that are necessary to specify these systems. The term *condition* evokes a useful association with conditions on agreement, which are factors that determine agreement in part but are not themselves realized directly (Corbett 2006: 176–205). Just as conditions on agreement do not directly specify an inflectional form, so also conditions on inflection do so only indirectly, as we shall see.

The place of conditions on inflection within the broader landscape of inflectional morphology is described in section 5.1, and is then demonstrated with a clear example in section 5.2. Then, in section 5.3, we present a subtly different example of a condition, which provides the basis for the analysis of these conditions, presented in section 5.4. This leads in turn to a typology of conditions, section 5.5, which is the core of the chapter. There are various observations to be made on this typology (section 5.6). Next, we review a particularly complex example of a condition in section 5.7, before sketching a higher-level typology (section 5.8), and concluding (section 5.9).

5.1 The Point: Conditions within Inflectional Morphology

Imagine we approach the inflectional morphology of a new language from scratch. We might set up a vast spreadsheet in which every token that we believe to be an inflected form is assigned to a unique cell, together with its context. We look for generalizations in two dimensions. The first is the lexical dimension, where we abstract lexemes from the data (abstraction of items with similar semantics, syntax, and phonology); traditionally these would be the columns of our spreadsheet. In the other dimension, the rows, we abstract morphosyntactic specifications (for instance, we might find that in contexts we treat as possessive there are forms which are morphologically identical to the subject of a nominalization, and hence we would label both as genitive). In the canonical world, the lexemes resulting from our analysis would be items with an unchanging stem, with all the variation located in the inflection. In the other dimension, we might find markers uniquely determined by the morphosyntactic specification, such that the same marker could be attached to any stem to yield a genitive. What we find in real systems is the need for additional statements, and that in both dimensions. First, having crystalized out the lexemes, we may find the need to make statements about groups of them that do not apply to

Table 5.1. *Verbal forms in Burmeso*

	assignment	'see'		'bite'	
		SG	PL	SG	PL
I	male	j-ihi-	s-ihi-	b-akwa-	t-akwa-
II	female, animate	g-ihi-	s-ihi-	n-akwa-	t-akwa-
III	miscellaneous	g-ihi-	j-ihi-	n-akwa-	b-akwa-
IV	mass nouns	j-ihi-	j-ihi-	b-akwa-	b-akwa-
V	banana, sago tree	j-ihi-	g-ihi-	b-akwa-	n-akwa-
VI	arrows, coconuts	g-ihi-	g-ihi-	n-akwa-	n-akwa-

Source: Donohue 2001: 100, 102.

others. These are likely to include part of speech classification (for syntactic purposes) and inflection classes (generalizations which are based on sets of lexemes for morphological purposes). But equally there are generalizations in the other dimension too, beyond what we can state once we have crystalized out the morphosyntactic specifications. It is these that are of interest here. We have presented this in abstractive terms (Blevins 2003: 742; Blevins 2006). We could view it in the alternative way, that of providing a description sufficient to construct all the forms. For our purposes the point of view adopted makes little difference and indeed, as Stump argues convincingly (2016: 257–264), treating these approaches as a strict choice creates a false dilemma.

To make all this more concrete, consider the fragment of the data for the verbal forms of Burmeso in Table 5.1 (entirely due to Donohue 2001: 100, 102). The prefixal gender-number markers mark agreement with the absolutive argument. We look for generalizations in two dimensions: those based on the lexeme and those starting from the morphosyntactic description. Generalizations in either dimension will be preferred when they cover more instances and permit fewer exceptions. Since the lexical and morphosyntactic dimensions are very different in range, a useful heuristic is the degree to which a generalization in one dimension cross-cuts distinctions in the other.

If we extend our fragment of Burmeso to cover the full verbal lexicon, we find a substantial number of verbs exactly like *-ihi-* 'see' and another equally large set like *-akwa-* 'bite'. Donohue (2001: 101) states explicitly that: '. . . there are no obvious semantic correlations for verbs which take the different sets of prefixes, and both sets of verbs are of approximately equal size.' Nor are there any phonological correlations relating to the form of the stem (Mark Donohue, p.c., 2 May 2013). Clearly, Burmeso requires lexical generalizations (inflectional classes)[1] since (1) the distribution of the affixes can be described only by reference to the particular lexeme (its membership in one of two inflection classes); and (2) the affixes are mutually predicting within the paradigm. The latter point demonstrates intra-paradigm structuring; this structuring supports lexicalist approaches, in a complementary way to the structuring demonstrated by suppletion-syncretism interaction (as found within Slovene nominal inflection, see Corbett & Fraser 1997, Brown & Hippisley 2012: 167–169).

Having established the need for lexical generalizations, we might hope for reasons of economy to eliminate generalizations in the other dimension. But we need look no further than the Burmeso data in (Table 5.1): we need to specify syncretisms (e.g., GENDER V SINGULAR = GENDER IV SINGULAR), since these cut across the inflection classes. This particular generalization refers to a featural specification as its antecedent and another as its consequent. Such conditions, in the clearest cases, are *orthogonal* to traditional paradigms; in some instances they may be exceptionless, but often they have interesting overrides.

We conclude that the minimal requirements for inflectional morphology are lexical generalizations and cross-cutting generalizations on inflection. This latter information is often excluded, by convention, from our representation when we analyse inflectional systems, or it may appear as accompanying notes. Here, in contrast, it is at the centre of attention.

5.2 A Clear Example: Count Nouns in Russian

We start from a clear example, then tease out its particular properties, in order to go on to propose a typology of these conditions. A fully explicit account of Russian noun inflection would include the example paradigms in Table 5.2. We could say that we have eight different paradigms here, since they are clearly distinct. However, that is not what is typically done. Rather we say that there are count and non-count nouns, the non-count having no plural (they are singularia tantum nouns). The condition involved here is that nouns falling below the number differentiability threshold on the Animacy Hierarchy (Corbett 2000: 54–88) have no plural. This approach, using a condition, makes good sense because it avoids two types of duplication. First, on the formal side, there are four major inflection classes in Russian, and each of them includes nouns that distinguish singular and plural and instances of singularia tantum (this

[1] In other languages we do find inflection classes which are predictable in part; such instances were discussed in Chapter 4.

Table 5.2. *Sample nouns in Russian*

	1	2	3	4	5	6	7	8
	'tram'	'bark(ing)'	'book'	'boredom'	'bone'	'pride'	'bog'	'milk'
NOM SG	tramvaj	laj	knig-a	skuk-a	kost´	gordost´	bolot-o	molok-o
ACC SG	tramvaj	laj	knig-u	skuk-u	kost´	gordost´	bolot-o	molok-o
GEN SG	tramvaj-a	laj-a	knig-i	skuk-i	kost-i	gordost-i	bolot-a	molok-a
DAT SG	tramvaj-u	laj-u	knig-e	skuk-e	kost-i	gordost-i	bolot-u	molok-u
INS SG	tramvaj-em	laj-em	knig-oj	skuk-oj	kost´-ju	gordost´-ju	bolot-om	molok-om
LOC SG	tramvaj-e	laj-e	knig-e	skuk-e	kost-i	gordost-i	bolot-e	molok-e
NOM PL	tramva-i		knig-i		kost-i		bolot-a	
ACC PL	tramva-i		knig-i		kost-i		bolot-a	
GEN PL	tramvaj-ev		knig		kost-ej		bolot	
DAT PL	tramvaj-am		knig-am		kostj-am		bolot-am	
INS PL	tramvaj-ami		knig-ami		kostj-ami		bolot-ami	
LOC PL	tramvaj-ax		knig-ax		kostj-ax		bolot-ax	

recalls the argument for recognizing deponency in section 4.4). The condition is orthogonal to the pattern of inflection classes. Second, on the semantic side, there would be a similar duplication. The semantic distinction goes across the different classes.[2] And while singularia tantum normally appear in the singular, there are instances of recategorization (Corbett 2000: 84–87), in which the plural is used; the noun is recategorized as a count noun (for instance, being recategorized as having a type reading), and then it has both singular and plural. Typically there is no doubt as to the form of this potential plural (see Zaliznjak 1967/2002: 57–58).[3] However, there can be overrides to such conditions; for instance, Zaliznjak points out (1967/2002: 57n38) that there are specific factors that make the plural of *moloko* 'milk' problematic – namely, the lack of obvious analogies and the existence of a competing lexeme (the plurale tantum noun *molóki* 'soft roe'); as a result, different plural forms are attested for it. There is cross-linguistic variation here, involving both the readiness with which recategorization is accepted and the readings available for particular nouns. We cannot assume the universality of the universal grinder and the universal packager (compare Pelletier 2012).

The example of count versus non-count nouns is one of the best cases for a condition. Specifically, the condition is that if a noun is very low on the Animacy Hierarchy (antecedent) it lacks a plural sub-paradigm (consequent). The condition is orthogonal to the inflection classes of Russian, whether we look at the forms or the meanings. This orthogonality means that the use of the condition avoids two types of duplication: (1) on the formal side, there are four major inflection classes and each of them includes count nouns and singularia tantum; (2) on the semantic side, similarly, the distinction goes across the different classes. We should ask what sort of condition this is. Its basis is clearly semantic, since its application depends on the semantic class of the noun. Its effect is on the shape of the paradigm: it specifies that certain cells (the plural

[2] The particular point at which we find the number differentiability threshold varies in interesting ways from language to language; Russian sets the boundary slightly higher than English (Corbett 2000: 80; see also Ljaševskaja 2004: 35–56). There is also the interesting phenomenon of 'number orientedness' (Polivanova 1983, Kulikov 2004: 127, Corbett 2012: 122–123) where nouns near the number differentiability divide have singular and plural forms available in principle but strongly favour one number value or the other. And there can be rapid change in the preferences of individual nouns (Gorbov 2014). There is also considerable interest in nouns which are pluralia tantum; see, for instance, Zaliznjak (1967/2002: 58–61, 75–80).

[3] Our thanks to Marina Chumakina and Alexander Krasovitsky for their help and intuitions here; *laj* 'barking', *gordost´* 'pride', and *moloko* 'milk', are taken from Zaliznjak's account. The other instance, *skuka* 'boredom', also occurs normally in the singular. There is no doubt about its plural forms (they would be as for *kniga* 'book'). There are a few examples of plural use in the Russian National Corpus, but they tend to be instances of word play.

ones) have a different status: they are not available to the same extent that the singular ones are. This is not a matter of particular inflections, but of the cells in the paradigm irrespective of their phonological form. Hence it would make no sense to treat this issue using morphemes. The fact that a slab of cells is involved suggests how we might build a typology, according to the basis of the condition and the type of its effect.

5.3 A Similar but Contrasting Example: Animacy in Russian

Russian shows a second very clear example, this time involving animacy and syncretism. We could propose eight main inflection classes for nouns, as shown in Table 5.3 (this table assumes the conflation of the eight types in Table 5.2 into four inflection classes). Though the evidence is not as abundant as for some other case values, we clearly need to postulate an accusative for Russian, since it has unique forms (just for nouns like *strana* 'country' and *djadja* 'uncle', and also for adjectives). Elsewhere, however, we find interesting patterns of syncretism. These are based on animacy, where grammatical animacy is close to semantic animacy; nouns denoting entities that live and move are animate (thus insects are animate but not plants).[4] Setting aside the nouns that have a unique accusative (which override the general conditions), the remainder are subject to the following condition:

> Nouns which denote animates and are masculine or plural have their accusative as the genitive.
> Other nouns have their accusative as the nominative.

This is a particularly good example of a condition. It cuts across all the inflection classes. There are even some neuter nouns which denote animates, and these are subject to the condition, having accusative-genitive syncretism in the plural, and accusative-nominative in the singular. The condition is interestingly different from the previous one since it affects not the cells available to particular items, but the patterns of syncretism which they show.

This pattern of syncretism has been discussed in various places. For Network Morphology, it is interesting to note that the analyses treat this condition quite differently from inflection classes; the condition is modeled using an evaluable path, which represents well the fact that the condition cuts across the different

[4] It appears that animacy in Russian is largely tied to the lexical semantics of the noun. It is not something that can readily be manipulated; however, just in numeral expressions there is some variability, based on referential status (see Mikaelian 2013, and the earlier valuable observations in Mel′čuk 1985).

Table 5.3. *The animacy condition in Russian*

	I		II		III		IV	
	‘factory’	‘boy’	‘country’	‘uncle’	‘bone’	‘mouse’	‘deed’	‘monster’
NOM SG	zavod	mal´čik	stran-a	djadj-a	kost´	myš´	del-o	čudovišč-e
ACC SG	zavod	mal´čik-a	stran-u	djadj-u	kost´	myš´	del-o	čudovišč-e
GEN SG	zavod-a	mal´čik-a	stran-i	djad-i	kost-i	myš-i	del-a	čudovišč-a
DAT SG	zavod-u	mal´čik-u	stran-e	djad-e	kost-i	myš-i	del-u	čudovišč-u
INS SG	zavod-om	mal´čik-om	stran-oj	djad-ej	kost´-ju	myš´-ju	del-om	čudovišč-em
LOC SG	zavod-e	mal´čik-e	stran-e	djad-e	kost-i	myš-i	del-e	čudovišč-e
NOM PL	zavod-y	mal´čik-i	stran-i	djad-i	kost-i	myš-i	del-a	čudovišč-a
ACC PL	zavod-y	mal´čik-ov	stran-i	djad-ej	kost-i	myš-ej	del-a	čudovišč
GEN PL	zavod-ov	mal´čik-ov	stran	djad-ej	kost-ej	myš-ej	del	čudovišč
DAT PL	zavod-am	mal´čik-am	stran-am	djadj-am	kostj-am	myš-am	del-am	čudovišč-am
INS PL	zavod-ami	mal´čik-ami	stran-ami	djadj-ami	kostj-ami	myš-ami	del-ami	čudovišč-ami
LOC PL	zavod-ax	mal´čik-ax	stran-ax	djadj-ax	kostj-ax	myš-ax	del-ax	čudovišč-ax

inflection classes (see Corbett & Fraser 1993: 129–132, Brown & Hippisley 2012: 80–83, 139–144). Müller (2007: 10–11) discusses Russian animacy from a slightly different perspective. The variation caused by animacy is predictable, and hence allows a reduction in the number of inflection classes, by grouping them into macro-paradigms (or macro-classes); this is a part of the agenda set by Carstairs' paradigm economy principle (1987) to constrain the number of inflection classes that can be found, given the inventory of markers (see also Blevins 2004). Müller himself takes a radical view, aiming to constrain the number of inflection classes, given only the inventory of markers and not having access to their associated feature specifications (Müller 2007: 18). More generally, the consideration of conditions links to the work of Dressler and colleagues on macro-classes, where the focus is on showing the relatedness of paradigms, in a hierarchical manner (Dressler 2003, Dressler, Kilani-Schoch, Gagarina, Pestal & Pöchtrager 2006; macro-classes are helpfully summarized in Gardani 2013: 26–27). The different point of view in the research reported here is that the emphasis is on the conditions themselves, rather than on seeing them as a justification for collapsing inflection classes. The question is what can cross-cut inflection classes and what is the typology of such conditions. This brings different items to the fore: if one is concentrating on collapsing classes, a means for combining two of them is ideal. If as here we wish to demonstrate a type of condition, the more classes which it cross-cuts the clearer the evidence for it.

5.4 Analysing Conditions

The Russian countability condition is convincing because it is orthogonal to the inflection classes. Its antecedent rests on the lexical semantics of the noun. Its consequent involves the abstract content of the paradigm (the number of cells, not their realizations). The animacy condition is similar but subtly different. Its antecedent again depends on the lexical semantics of the noun, but its consequent is a pattern of syncretism. It does not extend nor reduce the syntactic possibilities open to any given noun, but rather affects the pattern of realization. These conditions differ in their consequents; equally, it is not difficult to imagine conditions with different types of antecedent. This suggests a typology, built on possible antecedents of the condition and possible consequents. We discuss these in turn.

5.4.1 Possible Antecedents

We shall see that the antecedent of a condition on inflection may invoke four types of information. Semantic antecedents have already been provided in the

Russian examples. There are also syntactic (part of speech) antecedents: as we shall see, a particular condition may apply only to pronouns, for instance, and not to nouns. We also find conditions whose antecedents refer to morphological information, and others dependent on phonological information. All will be illustrated below.

5.4.2 Possible Consequents

The possible consequents of conditions require more detailed explanation. The different possibilities here take the discussion beyond that available in the major precursor – namely, Wurzel (1984/1989, 1990). We have already seen one difference: the Russian countability condition affects the shape of the paradigm, since a slab of cells is not available, whereas the animacy condition affects a pattern of syncretism. Yet other conditions (to be discussed later) have an impact more directly on forms. We may think of these consequents as different views of the paradigm, varying in terms of abstraction (Figure 5.1). The notion of abstraction is inherent in any inheritance model of paradigms; see also Williams (1994: 22).

At the most abstract level we have the content paradigm. This is the full inventory of content cells: these specify the contexts in which the lexeme can appear and its semantic interpretation. For instance, one cell might be specified as

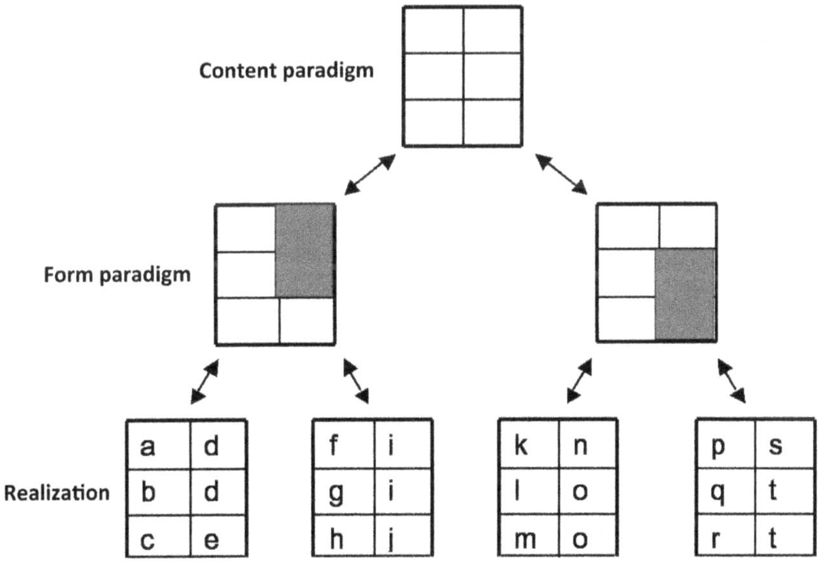

Figure 5.1. *Types of paradigm (compare Stump 2012)*

accusative plural, which determines its appropriate syntactic environments and its semantics (its grammatical meanings). Note that the syntactic context is not dependent on the actual inflected form of the lexeme, since syntax is morphology-free (see Zwicky 1992: 354–356, Corbett & Baerman 2006, and Corbett 2009b). If our lexemes in Figure 5.1 are nouns, with nominative, accusative, and genitive; singular and plural; each one has an accusative plural cell in its content paradigm.

Now consider the form paradigm. In the canonical world, the relation between content cells and their form correspondents is straightforwardly one-to-one. In the example in Figure 5.1, this holds true for most cells. There are two deviations, indicated by shading. In one, two cells (perhaps the nominative and accusative plural) share the same form (whatever that form is) and in the other there is a different pattern (the accusative and genitive plural share the same form). It is the form paradigm where such morphological generalizations are captured. The patterns here are syncretisms.

Finally we have realizations, the actual morphophonological expression, involving stems, affixes, and so on. Note that the first two paradigms showing realizations have different actual forms ('a' versus 'f' and so on), but both share the same form paradigm.

This account is inspired by Stump (2012), but does not match his division exactly; see further Stump (2016: 103–119). For our purposes, the key point is that lexemes may differ in realizations of particular morphosyntactic descriptions, yet share form paradigms. They may differ in form paradigms but share a content paradigm. And, of course, they may differ in content paradigms (e.g., some nouns have no plural). There are interesting problems involving a discrepancy between the content and the form paradigm – for instance where a small proportion of the relevant items have an additional cell (e.g., the Russian second locative, discussed under later section 5.5.11); for analysis of this general issue see Corbett (2010, 2014). Returning, then, to conditions, the consequent may involve the content paradigm, the form paradigm, or the realization.

5.5 The Typology of Conditions

We argue that there are four types of antecedent and three types of consequent; the combinations of these yield twelve theoretically possible types of condition (Table 5.4). These are labeled in the order in which the type of condition will be discussed. We adopt Bond's (2013: 24) notion of exploratory canonical typology, and set out to see which of these are attested. Each type (T1-T12) is considered in its own subsection.

Table 5.4. *Theoretically possible condition types*

antecedent	consequent		
	content paradigm	form paradigm	realization
semantic	Type 1	Type 2	Type 3
part of speech	Type 4	Type 5	Type 6
morphological	Type 7	Type 8	Type 9
phonological	Type 10	Type 11	Type 12

Table 5.5. *Diyari* pirta *'stick, tree' and* wirrawartanhi *'Farina'*

	'stick, tree'	'Farina' (place name)
ERG	pirtali	
NOM	pirta	
ACC	=NOM	
LOC	pirtanhi	wirrawartanhi
ALL	pirtaya	wirrawartaya
ABL	pirtandru	wirrawartandru
DAT	=ALL	

Source: Austin 1981/2013: 57, 58.

5.5.1 Antecedent: Semantic, Consequent: Content Paradigm (T1)

We have already seen an example of this type, the condition on count nouns in Russian (section 5.2). As a second example, in Diyari (a language of the north-east of South Australia, Austin 1981/2013), discusssed in section 4.3, the case system has seven values (Table 5.5). Most can be seen in *pirta* 'stick, tree'. The syncretic forms are distinct in other paradigms, hence it is fully justified to work with seven case values.

Unlike nouns like *pirta* 'stick, tree', location nominals (for spatial or temporal reference) have just three case values, as shown by the place name *wirrawartanhi* 'Farina'. This shows clearly that nouns of this type have a reduced content paradigm, preventing them from occurring in particular syntactic positions. Compare this with the syncretism in the common noun; there are different patterns for nouns of different semantic types (hence this would be a condition like those discussed in section 5.5.2, affecting the form paradigm), but these (non-locational) nouns do not lack any case values; they have a complete content paradigm.

5.5.2 Antecedent: Semantic, Consequent: Form Paradigm (T2)

This type was also illustrated earlier (section 5.3) in the discussion of Russian animacy. The other Slavonic languages show similar but interestingly different patterns of syncretism based on animacy (Huntley 1980).

For a rather different possible example consider the modal verbs of German (Wurzel 1989: 144–145). Modal verbs (such as *können* 'be able') have syncretism of the first and third persons singular in the present tense (*ich kann* 'I can' *er/sie kann* 'he/she can'); they are different from other verbs, which have distinct forms here. This special characteristic goes back to the preterit-present, and after various changes its range in Modern German is modal verbs only (plus *wissen* 'know'). It is Wurzel's argument about the development of the current situation which makes this example interesting: originally the modals and the verbs with preterit-presents overlapped only in part; subsequent changes (non-modals with preterit-presents gaining *-t* in the third person singular and modals which did not have the preterit-present losing the *-t*) have brought about a situation where the condition is almost exceptionless.

5.5.3 Antecedent: Semantic, Consequent: Realization (T3)

Polish provides a significant, if somewhat complex, example of this type of condition. It has four main inflection classes of nouns, and for three of these the default nominative plural inflection is *-i/-y*. However, if a noun denotes a male person, and it takes the default inflection, there is also a mutation of consonant where possible (Rothstein 1993: 698, 700). For example: *student* 'student', nominative plural *studenci*; *Norweg* 'Norwegian', nominative plural *Norwedzy*. Note that the condition involves persons: non-human animates do not qualify, as shown by *kot* 'cat', nominative plural *koty* with no mutation.[5] The alternations are shown in Table 5.6. We see from this that Polish has a complex system of alternation; the relevant set for the male personal nominative plural is that labeled 'alternation A'. This set applies not just in the context of interest here, and it contrasts with the second alternation set. More significantly, the alternation applies across the (relevant) inflection classes. Besides applying in the examples given, from the class that includes the majority of masculine personal nouns, it also applies to nouns denoting male humans that are in the inflection class whose members are mainly feminine nouns. The relevant nouns are those like *poeta* 'poet', with nominative

[5] Non-human animates show a pattern of syncretism, comparable to but more restricted to that of Russian (section 5.3), involving the form paradigm. The condition here, restricted to male humans, involves the realization paradigm.

Table 5.6. *Consonant alternations in Polish*

<table>
<tr><td></td><td>plain</td><td>alternation A</td><td>alternation B</td></tr>
<tr><td rowspan="2">velars</td><td>k, g</td><td>affricate c [ʦ], dz [ʣ]</td><td>alveopalatal cz [ʧ], ż [ʒ]</td></tr>
<tr><td>ch [x]</td><td>palatal ś [ɕ]</td><td>alveopalatal sz [ʃ]</td></tr>
<tr><td rowspan="3">others</td><td>r, ł [w]</td><td colspan="2">alveopalatal rz [ʒ], l</td></tr>
<tr><td>p, b, f, w, m</td><td colspan="2">palatalized p(i) [pʸ], b(i) [bʸ] f(i) [fʸ], w(i) [vʸ], m(i), [mʸ]</td></tr>
<tr><td>t, d, s, z, n</td><td colspan="2">palatal ć [ʨ], dź [ʥ], ś [ɕ], ź [ʑ], ń [ɲ]</td></tr>
</table>

Source: Based on Feldstein 2001: 25, discussed in Corbett & Baerman 2006: 241–242.

Table 5.7. *Morphology of Basque proper names*

	anthroponym	toponym
ABS	Esther	Gasteiz
ERG	Estherrek	Gasteizek
DAT	Estherri	Gasteizi
INS	Estherrez	Gasteizez
COM	Estherrekin	Gasteizekin
GEN	Estherren	Gasteizen
LOC	Estherrengan	Gasteizen
ABL	Estherrengandik	Gasteizetik
ALL	Estherrengana	Gasteiza

Source: Trask 1997: 96.

plural *poeci*, showing (like *student* 'student') the appropriate alternation from Table 5.6.

In Basque (Trask 1997: 95–96) animate nouns form their local cases (bolded in (Table 5.7)) differently from inanimates. This can be seen in the contrasting pair in Table 5.7. Although the inflection of these nouns is basically the same, we see that the animate noun shows an augment *-gan-* in the three local cases.

5.5.4 Antecedent: Part of Speech, Consequent: Content Paradigm (T4)

This type of condition may seem too obvious to include. After all, in German, for instance, only verbs inflect for person. In English, adverbs do not inflect. But, while in some familiar languages part of speech and inflectional behaviour line up fairly straightforwardly, there are many other languages where this is not so. These are interesting in their own right; moreover, they demonstrate that when we find implications from syntactic behaviour to inflection, these are not necessary but are of interest. A group that shows non-straightforward relations between part of speech and the content paradigm is the Daghestanian languages. We illustrate from Archi. In Archi it is relatively easy to demonstrate the parts of speech, using normal syntactic tests. Consider now the relation of part of speech to agreement in number and gender (and arguably in person too; see section 5.5). Almost all parts of speech can agree. We see this in the following example (Chumakina & Corbett 2015: 94):

(1) b-is o‹b›qˤa-t:-ib χ:ˤele
I/II.PL-1SG.GEN ‹I/II.PL›leave.PFV-ATTR-PL guest(I)[PL.ABS]
b-ez dit:a‹b›u e‹b›χni
I/II.PL-1SG.DAT soon‹I/II.PL› forget‹I/II.PL›PFV
'I soon forgot my guests, who had left.'

Agreement in Archi works on an ergative-absolutive basis. Agreement in (1) is controlled by the absolutive argument *χ:ˤele* 'guests'. The verb *eχmus* 'forget' follows the alignment pattern of verbs of perception: the experiencer stands in the dative, whereas the perceived entity is in the absolutive and controls the agreement. This verb has infixal marking, here <*b*>. The angle brackets follow the convention for infixes in the Leipzig glossing rules; information which can be inferred from the bare stem is indicated by square brackets. In addition in (1) every word is an agreement target (apart, naturally, from the controller *χ:ˤele* 'guests'). The participle *o‹b›qˤat:-ib* 'having left' agrees with the head noun *χ:ˤele* 'guests'; agreement in gender and number is marked by the infix, whereas the suffix marks number only (as in Table 5.10). The genitive pronoun *b-is* 'my' has a prefixal marker showing gender and number agreeing with its head noun. The first singular pronoun in the dative case, *b-ez*, agrees prefixally in gender and number with the absolutive argument *χ:ˤele* 'guests'. And the adverb *dit:a‹b›u* 'soon' agrees, also in gender and number, with the absolutive argument, this time with an infix. Other examples could be provided to show agreement of adjectives, the emphatic clitic, and a postposition.

So far we have seen that the range of items which can agree is impressively large. However, given the part of speech, this gives only a partial prediction about whether a given item will agree. Many do not (this is lexically rather than syntactically determined). The data in Table 5.8, from Chumakina & Corbett (2015: 95), derived from the Archi dictionary (Chumakina, Brown, Corbett, & Quilliam 2007) show the proportions of lexical items which agree. From this we see that within parts of speech only some items agree. For adverbs and postpositions it is a small minority which agree. Verbs are complex, with various interacting factors involved (see Chumakina & Corbett 2015: 102–115).

Table 5.8. *Agreeing lexical items in the Archi dictionary*

	total	agreeing	% agreeing
verbs	1248	399	32.0
adverbs	383	14	3.6
postpositions	34	1	2.9

Table 5.9. *An agreeing adverb in Archi:* k'ellej<t'>u *'entirely'*

<table>
<tr><th rowspan="2">gender</th><th colspan="2">number</th></tr>
<tr><th>SG</th><th>PL</th></tr>
<tr><td>I</td><td>k'ellej<w>u</td><td rowspan="2">k'ellej<b>u</td></tr>
<tr><td>II</td><td>k'ellej<r>u</td></tr>
<tr><td>III</td><td>k'ellej<b>u</td><td rowspan="2">k'ellej<t'>u</td></tr>
<tr><td>IV</td><td>k'ellej<t'>u</td></tr>
</table>

Personal pronouns are not included in Table 5.8, since it would be difficult to give meaningful figures. This is because some paradigm cells of some pronouns have a slot for agreement [as in example (1)].

There are two key points to retain. First, given the part of speech, we cannot uniquely predict an item's morphological behaviour in Archi. Thus languages in which there are absolute conditions based on part of speech classification (as noted earlier for German) represent one interesting possibility, but not the only one. Second, in languages like Archi, part of speech classification can be a powerful condition, but not an absolute one (as with adverbs which typically do not agree).

5.5.5 Antecedent: Part of Speech, Consequent: Form Paradigm (T5)

In the previous section we saw instances of gender/number agreement in Archi. This system includes patterns of syncretism, and these patterns are conditioned by the part of speech. For most parts of speech, the four genders distinguished in the singular are collapsed into two in the plural, as in this example of an adverb in Table 5.9. The same pattern of syncretism (involving forms which are similar though not identical to those in Table 5.9) is found in verbs, pronouns, and so on. In contrast, adjectives have a different pattern (Kibrik 1977: 114), as shown in Table 5.10:

Table 5.10. *The Archi adjective* haʁdu-t *'real, reliable'*

gender	number	
	SG	PL
I	haʁdu-(w)[a]	haʁd-ib
II	haʁdu-r	
III	haʁdu-b	
IV	haʁdu-t	

[a] The suffix [w] is not always pronounced, but it surfaces if there is a following vowel.

Thus, in Archi, the pattern of gender-number syncretism depends on the part of speech.

Further examples of this type include Lak (Daghestanian), where the first and second person pronouns have the ergative distinct from the genitive (though the ergative is syncretic with the absolutive); in contrast, nouns show the ergative syncretic with the genitive, and distinct from the absolutive (Kibrik 2003: 39, 40, 57). And in Archi, as partly illustrated already, gender/number agreement may be marked prefixally, infixally or suffixally; the choice is determined to a considerable degree by the part of speech (Chumakina & Corbett 2015: 100–102).

5.5.6 Antecedent: Part of Speech, Consequent: Realization (T6)

In Archi, various items show gender/number agreement by infixation (others show prefixation or suffixation). For those using infixes, there are two sets: the first is shown above in Table 5.9 for the adverb *dit:aw* 'early, soon'. This set is found with adverbs, the emphatic clitic, particles, and some pronouns. There is a second pattern, varying from the first in just one phonological form, used for gender IV singular and genders III and IV plural; here we find no realization (zero) in place of the <*t'*>. This second set is taken by verbs (see Chumakina & Corbett 2015: 96–99 for details).

The condition in Archi involves individual items according to their specific part of speech. In Imonda (a Papuan language) we find a condition that applies to items in different uses. There are certain verbs which may be used as full verb or as a classifier for another verb; they take different number markers for object agreement in the different roles (see Seiler 1985: 124–125).

5.5.7 Antecedent: Morphological, Consequent: Content Paradigm (T7)

We do not find examples of this type, and this is expected, since syntax is morphology-free (see section 5.4.2). Thus a morphological condition will not determine the content paradigm.

5.5.8 Antecedent: Morphological, Consequent: Form Paradigm (T8)

There are several examples of this general type. For one example, we return to the issue of deponency, introduced in section 4.4. The clearest instances are found in Latin (for a wider range of examples see Baerman, Corbett, Brown, & Hippisley 2007). We consider first the paradigm of a regular Latin verb in Table 5.11. Note that the passive forms are clearly distinguished from the active. Now compare the deponent verb in Table 5.12.

We have labelled the right column 'virtual'. It consists of cells which would be realized as given here, if they existed. They are struck through because they do not. The actual forms are active in use, but they have the appearance of passive forms (by comparison with the majority). For intransitive verbs, the effect is restricted to irregularity of form. But for a transitive verb, as here, the effect is more significant. Since the passive forms are "taken" for the active, the passive cannot be realized, and the verb is defective. (There are occasional exceptional examples in texts, but the basic point holds.) Thus, the morphological deponency of a transitive verb conditions its defectiveness with respect to the passive.

Table 5.11. *Partial paradigm of a regular Latin verb (*amāre *'love')*

	active	passive
1SG	amō	amor
2SG	amās	amāris
3SG	amat	amātur
1PL	amāmus	amāmur
2PL	amātis	amāminī
3PL	amant	amantur

Table 5.12. *Part of the paradigm of a deponent Latin verb (*mīror *'admire')*

	active	passive (virtual)
1SG	mīror	~~mīror~~
2SG	mīrāris	~~mīrāris~~
3SG	mīrātur	~~mīrātur~~
1PL	mīrāmur	~~mīrāmur~~
2PL	mīrāminī	~~mīrāminī~~
3PL	mīrantur	~~mīrantur~~

According to Wurzel (1989: 75) the Old High German instrumental would fit here as another example of this type of condition. He states that just in certain inflection classes the instrumental was available only in the singular. In other circumstances, a different construction with a preposition was required. Thus, Wurzel states that Old High German had a system of five case values side-by-side with one of four values (see also Luiten 2011). Thus the shape of the form paradigm depended on a morphological condition, the particular inflection class.

As a further example, in Archi, some verbs show gender/number agreement through prefixes and some through infixes (see Chumakina & Corbett 2015: 105, following Kibrik 1977). These are illustrated in Tables 5.13 and 5.14. In this type, the straightforward type for vowel-initial verbs, the agreement marker (in bold) is always prefixal. (Here we give only the gender III and IV forms, which are sufficient to indicate the position of the agreement marker. There is no overt IV gender singular exponent.) This means that when we look at the gender III forms in the imperfective, there is no conflict with the infixal imperfective marker <*r*>. Compare this with a verb taking infixal agreement, the basic option for consonant-initial verbs (Table 5.14).

Again all the agreement markers take the same position, namely infixal. This creates competition in the imperfective. The gender IV singular imperfective form *carχar* contrasts with the gender III singular: *cabχar*. Since the agreement marker of gender IV has no realization, this leaves the infixal position empty, and it is occupied by the imperfective exponent ‹*r*›. In gender III singular, the agreement marker ‹*b*› blocks the imperfective marker (though the imperfective has the overt realization -*r* in suffixal position).

Table 5.13. *The Archi verb* acu *'milk' (prefixal agreement)*

	perfective	finalis	imperfective	imperative
IV.SG	acu	aca-s	a‹r›ca-r	aca
III.SG	**b**-acu	**b**-aca-s	**b**-a‹r›ca-r	**b**-aca

Table 5.14. *The Archi verb* caχu *'throw' (infixal agreement)*

	perfective	finalis	imperfective	imperative
IV.SG	caχu	caχa-s	ca‹r›χa-r	caχa
III.SG	ca‹**b**›χu	ca‹**b**›χa-s	ca‹**b**›χa-r	ca‹**b**›χa

Table 5.15. *The Archi verb* akłu *'put through' (mixed prefixal and infixal agreement)*

	infixal		prefixal	
	perfective	finalis	imperfective	imperative
IV.SG	akłu	akła-s	a‹r›kła-r	akła
III.SG	a‹**b**›kłu	a‹**b**›kła-s	**b**-a‹r›kła-r	**b**-akła

These two types are the essential background for the lexically specified mixed type in Table 5.15, verbs which have prefixes for a part of the paradigm and infixes elsewhere. The contrasting forms shown here allow us to see what is going on. The gender III singular forms of the imperfective and imperative show that this verb has a perfectly good prefixal agreement slot available. However, this is not the one used in the perfective and finalis, where the infixal position is used; this demonstrates that Archi has 'frivolous' infixation (Yu 2007: 41–42). That is to say, it is not phonologically conditioned. In the imperfective, the <*r*> infix of the imperfective marker appears, and gender/number marking is prefixal in this form; it is also prefixal in the imperative, where there is no similar competition.

The key point for us is that the combinations are not random; rather the morphological pattern of stems determines the distribution of affixation in the mixed verbs. All verbs of the mixed type, just like the one in Table 5.15, have infixal agreement in the perfective and finalis (and all the forms based on these stems) and prefixal agreement in the imperfective and imperative (and forms based on these stems). Thus the main stems are paired, and this pairing splits the verb paradigm and determines whether agreement will be prefixal or infixal in the mixed type.

5.5.9 Antecedent: Morphological, Consequent: Realization (T9)

German nouns which form the plural with *-er* (with umlaut, if the stem vowel is subject to umlaut), will have the genitive singular in *-(e)s*. For instance, *Kalb* 'calf', has the plural *Kälb-er*, and the genitive singular *Kalb-es*, in accordance with the condition. Note that the implication goes in this direction, and not from the genitive to the plural. Thus *Hund* 'dog' has the genitive singular *Hund-es*, but its plural is *Hund-e* (Wurzel 1989: 114). Similarly for German, Andrew Spencer points out (personal communication) that if a verb has a strong past participle it follows that it will have a strong past tense.

Table 5.16. *Verb forms of Võro with stems ending in a short vowel*

	'shear'	'be born'	'revile'	'scurry'
PRS 3SG	pügä	sünnü-s	`sõima-s	sipa
PST 1	`pöi	`sündü	`sõim-ssi	siba-si
INF	pükä-ʔ	`sündü-daʔ	sõima-daʔ	sipa-ʔ
GER	pükä-n	`sündü-den	sõima-ten	sipa-n
JUSS	pügä-guʔ	`sündü-guʔ	sõima-kuʔ	siba-guʔ

Source: Baerman 2014, based on Iva 2007.

A more complex example is provided by the verb forms of Võro, a Finnic language of Estonia. The possibilities for verb inflection are numerous and complex; let us set aside the instances in which forms can be predicted from phonological information, as with consonant final stems and those ending in a long vowel or a diphthong, and confine ourselves to those which end in a short vowel, illustrated in Table 5.16.

Consider the different inflections from verb to verb. There is no obvious phonological motivation for the differences. However, there is a subtle morphological prediction. Võro verbs may have weak and strong stems (recall the discussion of strong and weak in section 2.2); in Table 5.16, the cells with weak stems are shaded, and the remaining cells have strong stems. To determine whether a stem is strong or weak, one cannot simply inspect a stem and know. Rather, it is only by comparing the two that one can see which is weak and which is strong. Furthermore, knowing which stem is found in a particular cell is not sufficient for predicting the inflection in that cell. The key information is the *pattern* of weak and strong cells, and this pattern allows a prediction of the set of inflections. (Even then, there are additional stem alternation patterns that remain underdetermined by these predictions; in such instances the suffixes are lexically specified.) Thus, this is clearly a morphological condition, whose antecedent is the pattern of weak and strong stems.

Conditions may be bidirectional. Thus, Spencer (1991: 226) notes an interesting subclass of Czech nouns with 'soft' stems. In particular, there are nouns, several of which denote the young of animals, which take the augment *-et-* in the oblique cases of the singular. These nouns also take the augment *-at-* throughout the plural, as shown in Table 5.17. This is bi-conditional: nouns with the augment in the plural will also have an augment in the singular (oblique case values).

There is a comparable but more complex example of a condition based on stem augments in Russian, involving ethonyms (Baerman 2014: 7). There

Table 5.17. *Czech* jehně *'lamb'*

	SINGULAR	PLURAL
NOM/VOC/ACC	jehn-ě	jehň-at-a
GEN	jehn-ět-e	jehň-at
DAT	jehn-ět-i	jehň-at-ům
INS	jehn-ět-em	jehň-at-y
LOC	jehn-ět-i	jehň-at-ech

Orthographic note: *ɲ* is spelled as *ň*, but the combination *ň* plus *e* is spelled *ně*.
Source: Short 1993a: 470.

are two basic patterns. In the first type, the augment is found in the singular only (e.g., *armjan-in* 'Armenian', plural *armjan-e* 'Armenians'). These nouns have the special nominative plural affix *-e*, in contrast to the default *-y*. In the other basic pattern, the *-in-* is found throughout the paradigm, and the default nominative *-y* occurs (e.g., *gruz-in* 'Georgian', plural *gruzin-y*). The condition appears to be purely morphological. However, there is an additional point. The stem (minus the augment) must end in *-n*. Hence *tatar-in* 'Tatar' has the plural *tatar-y* 'Tatars'. There is no augment in the plural, but the basic stem ends in *-r*, hence the default plural form. Thus, this condition has both morphological and phonological antecedents.

5.5.10 Antecedent: Phonological, Consequent: Content Paradigm (T10)

No examples of this type have been found. Yet this is precisely the gap we might have expected; if we did find phonological conditions on the content paradigm, these would go counter to the claim that syntax is phonology-free (Zwicky & Pullum 1983).

5.5.11 Antecedent: Phonological, Consequent: Form Paradigm (T11)

Russian provides an instance of this type of condition. Russian has an imperfective gerund, as in *čitaja* '(while) reading', *govorja* '(while) speaking'. However, for verbs with an asyllabic stem (which would give a monosyllabic form), the imperfective gerund is missing (or awkward). For instance, **vrja* 'lying, telling untruths', or *?ždja* 'waiting'. As the latter example indicates, there is variability between verbs (and also between speakers). For information see Švedova (1980: 673), and Timberlake (2004: 96); Timberlake states that monosyllabic forms are awkward, but offers a textual example of *rvja* 'tearing' to demonstrate that they are possible. This type of condition involves defectiveness (Baerman, Corbett, & Brown 2010, Sims 2015, Stump 2016: 157–168).

A less clear example, also from Russian, involves nouns with the 'second locative'. The ordinary locative is straightforward; it is also known as the prepositional case, since it occurs only with prepositions, as in *v gorode* 'in town', *o knige* 'about the book'. There is a second locative, for a minority of nouns (not including the two just given, which have the same form for both). A noun with the second locative is *sad* 'garden', as shown in these contrasting examples:

(2) o sad-e
about garden-SG.LOC
'about the garden' (on the subject of the garden)

(3) v sad-u
in garden-LOC2
'in the garden'

This distinct form, the second locative, is restricted in several ways. It occurs with only two prepositions (*v* 'in' and *na* 'on') and mainly if they have their basic locative or temporal sense. The second locative is distinct for relatively small groups of nouns (currently under 150 in total). However, these nouns are of different inflection classes: most are from the same class as *sad* 'garden', but some are from another class, for example *v krovi* 'in blood' (covered in blood), where the distinction is one of stress only. Brown (2007) discusses the development of the second locative, and gives further references, while Plungian (2002) concentrates on its semantics. Recent examples can be found in Sakhno (2011).

How is this to be handled? Brown treats it as a 'structured' case value (1998: 198–200), discussed in Corbett (2012: 206–209). In his Network Morphology analysis, the default realization of the second locative will be as the normal locative, and this default is overridden for the relatively few nouns with a distinct second locative. From the perspective of our typology, the content paradigm will include the second locative, since the syntax can require this case value, but the majority of nouns will have the first and second locative forms identical in the form paradigm. We are interested in the nouns which have a separate second locative in the form paradigm, and we find one particularly relevant condition (see Brown 2007: 66):

For a noun to have a second locative, it must have a monosyllabic stem.

Brown points out that there are a few exceptions, but that these are motivated (typically they are compounds based on monosyllabic stems). Note, however, that this is a necessary condition (not forgetting the few exceptions just mentioned) but not a sufficient one.

A clear-cut example is provided by Stump (2015: 127, based on Tucker 1940). The Nilo-Saharan language Moru, spoken in Sudan, has three inflection classes of verbs. Assignment to these classes is phonological: monosyllabic verbs form one class, disyllabic verbs form two more, depending on whether they are vowel initial or consonant initial.

5.5.12 Antecedent: Phonological, Consequent: Realization (T12)

A relatively simple condition of this type is that Russian nouns whose stems end in a vowel cannot inflect (Worth 1966, Corbett 1982: 218). For instance, *kenguru* 'kangaroo', *gnu* 'gnu', *taksi* 'taxi' and *kino* 'cinema' are all uninflecting, as noted in section 2.4. Note, however, that they are not defective. They can appear in all the syntactic slots, for all the case/number combinations, but they do not change in form.

Similarly, in Archi, simple dynamic verbs with the perfective stem in *b-* do not inflect (Kibrik 1977: 80), even infixally (Chumakina & Corbett 2015: 104). It would appear that this is because this form has the appearance of one with an initial gender/number marker. However, those with initial *d-*, which is also equivalent to a gender/number marker, give no prediction (Chumakina & Corbett 2015: 104).

A condition involving straightforward affixation is found in the West Slavonic language Slovak. In what we shall call inflection class I, which is the inflection class including nouns like *chlap* 'fellow' and *dub* 'oak', Short (1993b: 541) points out this condition for the instrumental plural: '*-mi* is used after single stem-final consonants except *-m*', while . . . *-ami* is used 'after *-m* or consonantal clusters . . .'. Some examples are given in Table 5.18. Although this condition is attractively straightforward, it appears to apply only to this paradigm; the 'choice' is between two affixes (*-mi* and *-ami*, which are found in other inflection classes). However, *-ami* is the default inflection here, so we could generalize the condition along these lines (simplifying a little):

> If a noun is of inflection class I
> *and* its stem ends in a single consonant other than *m*
> the instrumental plural consists of the plural stem + *-mi*.
> *Otherwise* the instrumental plural consists of the plural stem + *-ami*.

Table 5.18. *Conditions on the instrumental plural in Slovak*

	NOM SG	INS PL	reason
'fellow'	chlap	chlap-mi	single stem-final consonant, not *-m*
'oak'	dub	dub-mi	single stem-final consonant, not *-m*
'house'	dom	dom-ami	stem-final *-m*
'bridge'	most	most-ami	stem-final cluster

For further examples, and discussion, see Paster (2015). While discussing phonological conditions, it is important to distinguish these from phonological rules. Phonological rules are synchronically motivated. Conditions can have antecedents stated in phonological terms, but they are not synchronically motivated. This is stressed in Baerman (2014: 10), distinguishing what is phonologically motivated from what is phonologically predictable (the latter being our concern here); for previous references to the distinction, see Anderson (2008), Carstairs (1987: 21; 1988; 1990), and for earlier phonological rules becoming morphological, see Anderson (1992: 343).

5.6 Observations on the Typology of Conditions

Given the typology laid out in (4), we might reasonably have expected to find types which are not attested. However, the typology proves remarkably complete. Where possible, we have given clear and unambiguous examples, but there are also some which are less clear. For example, the Slovak condition based on the stem-final consonant (section 5.12) is straightforward in terms of antecedent and consequent, but it does not cut across the different inflection classes as fully as a canonical condition does.

We also find examples of conditions with various overrides. Thus, a condition based on semantic animacy may apply also to some items that are semantically inanimate (giving a mismatch between semantic and grammatical animacy), as with Polish nouns like *walc* 'waltz' and *pech* 'bad luck' (see, for instance, Swan 1988). Or in Russian, the part of speech of an item typically gives clear predictions as to its inflectional morphology, but there are items like *stolovaja* 'dining room', which function as nouns but have the inflectional morphology appropriate to an adjective (see Spencer 2013: 122–123 for an analysis of this mismatch, in the context of a general typology of lexical relatedness). Also, in Russian there are conditions based on the stem-final consonant, in particular whether it is palatalized; these deal in what the morphology "considers" to be palatalized rather than in synchronic palatalization (see Baerman 2001 for further complications here). For a Distributed-Morphology perspective on conditioning factors, see Embick & Noyer (2007: 299 n15), who point to Bobaljik (2000) and Embick (2003).

A further refinement is that conditions can be complex – involving more than one antecedent. David Gil (personal communication) suggested that all the proposed conditions have an implicit part of speech antecedent. This is not quite right. For instance, the Russian animacy condition (section 5.3) applies to all items with case (straightforwardly to nouns and adjectives, in more limited ways to pronouns and numerals). And as we stressed in section 5.5.4, although some of the more familiar and well-studied languages have fairly monolithic parts of speech, with morphology and syntax aligning well, there are Daghestanian languages, like Archi, where the part of speech gives only limited predictions about inflectional morphology. We discuss complex conditions, largely through a case study, in section 5.7.

5.7 Complex Conditions: The Serbo-Croat[6] Augment

We have already seen that conditions may involve more than simple antecedents: we noted, for instance, the Polish example in section 5.5.3 involved nouns which (1) denote a male person and (2) take the default inflection. Indeed, Wurzel (1989: 114) states that there can be relationships "between combinations of extramorphological and morphological properties on the one hand and morphological properties on the other." Here we present the essentials of a fascinating and complex condition. Consider these contrasting paradigms in Table 5.19.

Table 5.19. *The Serbo-Croat augment (in inflection class I)*

	'window'	'city'
NOM SG	prozor	grad
VOC SG	prozor-e	grad-e
ACC SG	prozor	grad
GEN SG	prozor-a	grad-a
DAT SG	prozor-u	grad-u
INS SG	prozor-om	grad-om
LOC SG	prozor-u	grad-u
NOM/VOC PL	prozor-i	grad-ov-i
ACC PL	prozor-e	grad-ov-e
GEN PL	prozor-a	grad-ov-a
DAT PL	prozor-ima	grad-ov-ima
INS PL	prozor-ima	grad-ov-ima
LOC PL	prozor-ima	grad-ov-ima

[6] We use 'Serbo-Croat' here to cover Bosnian, Croatian, Montenegrin and Serbian (Corbett & Browne 2009). For recent discussion see Bugarski (2012) and von Waldenfels & Eder (2016). Note especially that the phenomenon is found in Štokavian, the dialect basis of all these variants; the other variants are dramatically different. Kajkavian (spoken in the north of Croatia) did not develop the long plural, whereas Čakavian, spoken to the west, particularly on the islands, began the development but lost it by the sixteenth century (Orlandi 1963: 8–11).

Nouns of the first inflection class in Serbo-Croat may have an augment right through the plural paradigm (Browne 1993: 319–320, following Ivić 1972: 112–116). While nouns like *prozor* 'window', have a set of plural inflections (the 'short plural') which are comparable to those of the singular, other nouns have the augment *-ov-*, as in our example *grad* 'city', with the nominative plural *grad-ov-i* (the 'long plural'). We need a condition to specify which is found with a given noun.

A first attempt, which covers much of the data, would be:

> if a noun is of inflection class I, and its stem is monosyllabic
> it will take the augment *-ov-* in the plural.

Thus, the antecedent is both morphological (inflection class) and phonological. The phonological part involves syllable counting (see, for instance, Franks 1990). What counts as monosyllabic is interesting. Those who have driven along the Croatian coastal road are likely to have seen several signs offering *autodijelovi* 'car parts' (showing the augment). The corresponding singular is *autodio*, and treating that as monosyllabic will require a somewhat rubbery definition. First, it is a compound, and the syllable count is of the final element (thus *auto-* is discounted). There is a general rule that final *-l* becomes *-o*, and this *-o* does not count as a vowel for the purpose of the rule. Even so the correspondence between singular *autodio* and plural *autodijelovi* needs a further comment. The vowel *ě* (jat´), when long, gave rise to *ije* in the Ijekavian variety. This resulting *ije* counts as a single vowel for the purpose of the rule; the impact of the Ikavian variety is suggested as the cause for instances of *-i-* from *-ě-*, as in the singular (see Vukušić, Zoričić, & Grasselli-Vukušić 2007: 23 and references there). The basic point is that 'monosyllabic' is not straightforward. And borrowings with non-native vocalic groups can be treated as monosyllabic, for instance *faul* 'foul', plural *faulovi* (Nikolić 2013: 301).

More seriously, there are counterexamples in both directions. Some of these are predictable. Let us start with monosyllabics which have the short plural (do not take the augment *-ov-*). Ethnonyms are a large group which behave in this way: *Rus* ~ *Rusi* 'Russian(s)', *Kurd* ~ *Kurdi* 'Kurd(s)'. This generalization holds even when other phenomena are involved. There may be a mutation of consonant: *Grk* ~ *Grci* 'Greek(s)', *Čex* ~ *Česi* 'Czech(s)', or an augment in the singular but not in the plural and a mutation: *Turčin* ~ *Turci*. Finally, some examples (not only ethnonyms) have a "fleeting vowel" in the nominative singular but not in the plural: *Nemac* ~ *Nemci* 'German(s)'. Nouns denoting measures behave similarly in having the short plural, for example: *volt* ~ *volti* 'volt(s)', *jard* ~ *jardi* 'yard(s)'. These items are borrowings, but this is not in itself a determining factor; thus, the borrowing *film* 'film' has the expected plural *filmovi*. As noted later, it is significant that *sat* 'hour, watch, clock', when used as a unit of measurement normally has the plural *sati* 'hours' as opposed to *satovi* 'watches, clocks'. The complexity of the conditioning is now becoming clear: we have had a condition based on a morphological class, with a phonological condition (syllable counting) and semantic conditions (ethnonyms and nouns denoting measures). There is a link between the semantic groups: it is suggested that the nouns involved are used more in the plural than in the singular. Thus Pešikan (1956: 275) suggested that nouns like *gub* ~ *gubi* 'lip(s)', *mrav* ~ *mravi* 'ant(s)', are more frequently used in the plural, and these have the short plural. This argument carries over to ethnonyms and units of measure (see also Ivić 1972: 113–114 and Nikolić 2013: 310n163 for discussion).

Just as there are nouns with monosyllabic stems which take the short plural, so there are also nouns with bisyllabic stems which take the long plural. Typically, such nouns also allow the short plural; there are different ratios of long to short plural, but there are some nouns for which the long plural is much the more frequent choice (Nikolić 2013: 14). These include: *slučaj* ~ *slučajevi* 'chance(s), case(s)', *golub* ~ *golubovi* 'pigeon(s)', *plamen* ~ *plamenovi* 'flame(s)'. With nouns of more than two syllables, the long plural can be found in those which are compounds whose second element has the long plural (as discussed earlier with *autodio* 'car part'); see Klaić (1953) for discussion of nouns with longer stems.

There is a good deal of variation in the system, with some nouns taking both long and short plurals with different degrees of preference. However, this variability is found much more with the bisyllabic stems than the monosyllabic ones. For data from a remarkably large text count, involving over 330 literary works (mainly post-1950 prose) and some 30 additional sources see Nikolić (2013); Samardžija (1988) gives statistics on 972 nouns taken from their dictionary entries. Given the variability, it is not surprising that there are instances of stylistic and semantic differentiation; these are considered particularly by Šipka (1999). A standard example is: *sat* meaning both 'watch, clock' and 'hour', which has the plurals *satovi* 'watches, clocks' and *sati* 'hours'. This would seem to fit the earlier generalization well, in that we expect the singular to be more frequent in the meaning 'watch', while with 'hour' the plural would be expected to be the more frequent (Nikolić 2013: 289).

The fact that there is greater variability in the bisyllabic suggests a diachronic scenario, where the long plural invades first the monosyllabic stems, and then later moves on to the bisyllabic ones. And indeed, there is an interesting history to this condition. It goes back to the paradigm of **ŭ*-stem nouns in Common

Table 5.20. *The rise of the long plural in monosyllabic nouns*

author	dates	plural forms (*zakon* inflection class)	long plural	% long plural
B. Zuzzeri	1682–1762	375	73	19.5%
A. Kačić-Miošić	1704–1760	1091	426	39.0%
M. A. Reljković	1732–1788	472	248	52.5%
D. Obradović	1742–1811	470	281	59.8%

Source: Orlandi 1963: 20–21.

Slavonic. The number of such nouns has been considerably discussed. The number suggested is usually quite low (under twenty), but Orr (1996) reviews the literature and argues there may have been many more. He lists the following six (1996: 317) as those which are universally accepted: **domŭ-* 'house', **medŭ-* 'honey', **polŭ-* 'half, sex', **synŭ-* 'son', **volŭ-* 'ox', **vĭrxŭ-* 'top, summit'.

The typical development in Slavonic languages is that members of this small class join other inflection classes, often contributing an inflectional form (or a variant form). Thus the *-owi* of the Polish dative singular and the Russian genitive plural in *-ov* go back to the old **ŭ*-stems. The Russian second locative discussed in section 5.5.11 also goes back to the **ŭ*-stems (see Schenker 1993: 87 for the forms). The plural with *-ov-* is found in different languages, most notably in Serbo-Croat's close relative Slovene, which has it also in the dual, but for a restricted number of nouns, often as a variant. It is also found for monosyllabics in the other South Slavonic languages, Bulgarian and Macedonian; in these two languages, the loss of case distinctions means that we have essentially a singular-plural opposition. Taking all this together, this means that the Serbo-Croat situation is unique in the Slavonic family, in that the plural augment *-ov-* has made the greatest progress in terms of the number of nouns affected, including polysyllabics, and in going right through the plural. Nikolić (2013) shows that in the modern language there are hundreds of nouns, which have *only* the long plural (and there are many in addition which allow both plurals). The rise of the long plural is documented by Orlandi (1963), and the reader is directed there for the detail. As one part of the story, however, consider the data in Table 5.20, from the period when the long plural made a dramatic advance.

All the writers included had the Štokavian dialect, but with different varieties as noted. They include two prose writers, B. Zuzzeri (Cuceri, 1682–1762,

Jekavian) and D. Obradović (1742–1811, Ekavian), and two authors who wrote poetry and prose, A. Kačić-Miošić (1704–1760, Ikavian) and M. A. Rejković (1732–1788, Ikavian). Zuzzeri represents the conservative usage of Ragusa (Dubrovnik), while Kačić, a Dalmatian writer, uses many more long plurals. The literary language of Reljković and Obradović shows another step forward, with a preponderance of long plurals. Contemporary usage shows further development along the same lines. Though not included in the table, there was also a clear development in this period with polysyllabic stems: Zuzzeri used the long plural for just two such nouns, whereas the other three have the long plural for a wider range of nouns (Orlandi 1963: 211).

We have seen some indication of how a condition can arise from a small number of instances and extend dramatically over time. The original small group of *ŭ-stem nouns had monosyllabic stems, and the long plural spread first in this domain. It is now dominant in the monosyllabic nouns, and is extending to the others. Let us now consider the status of this condition. Recall that there is an interaction of antecedents within the condition, including inflection class. This is apparently a poor instance of a condition because it affects one inflection class, rather than being orthogonal to that classification. In other respects, however, its credentials as a condition are better. First, it affects not the realization of a single cell but rather a 'slab' – namely the whole plural subparadigm. Second, it cuts across a second phenomenon. Although we have referred to the 'long plural' and the 'augment', there are actually two possibilities here, *-ov-* and *-ev-* (we saw the latter in the example *slučajevi* 'chances' above). The basic criterion is that *-ev-* is found after palatals and *c*, but the distribution is not fully predictable, and there are various nouns which allow both. For early discussion, see two journal contributions (Vuković 1949 and Milanović 1949) and the editorial response (Uredništvo 1949). For fuller accounts see Ivić (1972: 112–113, and especially Nikolić (2013: 305–309). The relevance is that the antecedents for the condition on the use of the augment cross-cut the factors which determine the variant to be used (that is, the choice of long or short plural appears to be independent of the form of the augment that is used).

Given that the condition involves a single inflection class, it can be asked why the phenomenon is treated as a condition, rather than by postulating the existence of two inflection classes. We think that a plausible route for the rise of inflection classes is precisely through the breakdown of conditions, but we suggest that this has not yet happened in Serbo-Croat. In brief, there are two arguments. First, there are various minor variants of the singular inflections within the inflection class (largely but not fully based on phonology), and typically the long and short plurals are available across these variants. Second, once a noun takes the long plural, the plural inflections will be regular (the forms given in (Table 5.19)), whereas there can be irregularity in the short plural (in the genitive plural, see Nikolić 2013: 286–288, 311). If nouns with the long plural were treated as belonging to a separate inflection class, we would introduce redundancy, in that their plural inflections would match exactly those of the regular members of the inflection class with the short plural.

The condition involving the Serbo-Croat augment is complex and interesting; its rise is dramatically different from the story of the other Slavonic languages, which had, in a sense, the same starting conditions. It demonstrates that while we concentrated initially on clearer instances of conditions, there are also instances with multiple antecedents, and numerous items allowing variation in the outcome. A similarly complex story involves noun plurals in Burushaski, for which see Bye's interesting account, which also includes a diachronic dimension (Bye 2015).

5.8 Types of System

Our analysis suggests a second, higher-level typology, based on the relative importance of inflection classes versus conditions on inflection in a given language. The most cogent case for conditions comes from systems in which the condition is orthogonal to a well-justified classification in the other 'dimension', as in our Russian examples. However, there are other possibilities. An extreme system is the verbal morphology of Burmeso (Donohue 2001), with fully discrete inflection classes, which partition the verbal inventory and leave no need for conditions; however, as we saw in section 5.1, there are clear morphomic (morphology internal) patterns in Burmeso. Although we do not need conditions to account for the forms in Burmeso, we can hardly ignore the patterns which call for a condition. Then we find systems like that of Russian which allow for inflection classes *and* for orthogonal conditions (with differing relative weights). At the other extreme we find Diyari, whose nouns have been described (Austin 1981/2013) in terms of semantic and phonological conditions, with no place for inflection classes.

5.9 Conclusion

Conditions on inflection are an essential part of inflectional description. It is worth reflecting, therefore, on the practical question of how conditions should be represented when we present morphological data. We have some informal conventions for presenting paradigms (see Corbett 2013 for suggestions on formalizing these somewhat), but conditions are too often treated as minor addenda to the presentation of inflection classes, or even relegated to footnotes. One means to integrate conditions into the representation is that used in Table 5.3, where the range of the condition and its cross-cutting effect are represented in a simple graphical fashion.

The typology of conditions established in this chapter is quite rich and complete. We might have expected to find semantic antecedents for consequents involving the content paradigm, and phonological antecedents for consequents involving the realization. But we also find the less obvious intermediate types. That is, given four types of antecedent and three types of consequent, we find ten of the logically possible types of condition, and we did not expect to find the remaining two (they would have run counter to the principles of morphology-free and phonology-free syntax). In addition, we find complex conditions, as illustrated by the Serbo-Croat augment. This typology opens the way for us to investigate the distribution and relative frequency of the different types. We should also investigate further the extreme language types discussed in section 5.8, currently represented by Burmeso and Diyari. We should research the diachrony of these systems, with an initial hypothesis being that the natural progression is from condition-based systems to inflection class systems.

6 *Paradigm Structure*

When we speak of inflection class, we have in mind more than just individual differences in form. Inflection generates multiple forms, so inflection classes are sets of words that differ at multiple points in their paradigm. The way these differences are spread over the paradigm constitutes a typological parameter in its own right, one which has figured prominently in recent developments in morphological theory. In section 6.1 we look at the variety of configurations possible under a system of allomorphic classes, comparing that in section 6.2 to the equivalents in systems that are defined distributionally. In section 6.3 we look at how the various configurations can be combined, which gives us a sense of how actually attested systems are constituted. In section 6.4 we step back and look at what happens when a language has morphologically distinct parallel systems of inflectional marking, each of which displays inflection class distinctions.

6.1 Allomorphic Classes

Consider first the typological extremes. In what we can call a canonical inflection class system, each point in the paradigm differs from every other (Corbett 2009a, 2015: 162, Stump 2016: 31–42, 84–90). This means that each class is in effect a self-contained entity, as with the Czech non-past verb forms in Table 6.1. Contrast this with the verbal paradigms in Table 6.2 from Seri. The paradigms differ in their plural subject forms (see section 6.2.1 for a discussion of the morphosyntactic values behind this paradigm). The singular event forms have the suffixes *-t*, *-tax*, or *-tox*, the plural event suffixes have *-tołka* or *-tam*. But instead of forming self-contained classes, the allomorphs cross-classify: each singular-event suffix can occur with each plural-event suffix.

Clearly, these are classes in a different sense from those in Czech. Rather than being made up of a unique SET of exponents, they are made up of a unique

Table 6.1. *Canonical inflection classes in Czech verbs*

	'make'	'judge'	'lead'
1SG	děl-ám	soud-ím	ved-u
2SG	děl-áš	soud-íš	ved-eš
3SG	děl-á	soud-í	ved-e
1PL	děl-áme	soud-íme	ved-eme
2PL	děl-áte	soud-íte	ved-ete
3PL	děl-ají	soud-í	ved-ou

Source: Janda & Townsend 2002: 35–36.

Table 6.2. *Cross-classifying affixes in Seri: third person dependent realis forms*

		'go'	'go (to get)'	'make a scratch or incision'	'cause to be'	'throw (stones) at'	'bring cactus fruit w/ thorns'
SG SBJ	SG EVENT	kontíta	itóːi	itʃéxe	itaʔáː	ittím	taʔíkʷs
SG SBJ	PL EVENT	kontíta-tim	itóːi-tim	itʃéxe-tim	itaʔáː-tim	ittím-tim	taʔíkos-im
PL SBJ	SG EVENT	kontíta-t	itóːi-t	itʃéxe-tax	itaʔáː-tax	ittím-tox	taʔíkos-ox
PL SBJ	PL EVENT	kontíta-tołka	itóaː-tam	itʃéxe-tołka	itaʔáː-tam	ittím-tołka	taʔíkos-am

Note: Suffix-initial /t/ is retained (1) after vowels, and (2) in an immediately post-tonic syllable after a single consonant; elsewhere it is deleted (Marlett 1981, 2015).
Source: Moser & Marlett 2010.

combination of exponents. This sort of contrast has driven a large body of research into the structure of inflectional paradigms. Systems like Czech have a deterministic implicational structure: any single form unambiguously implies the remainder of the paradigm. Systems like Seri, on the other hand, have no implicational structure: no form implies anything about what goes on in the rest of the paradigm. Unsurprisingly, most attested systems fall somewhere in between these two extremes, and if we look at real systems in their entirety, we may well despair at ever getting a typological overview.

To lend some clarity to our discussion let us step back and think in abstract terms. As a point of departure, say that we have a system with four inflection classes. How can these classes actually be constituted? We start in Table 6.3 with an abstract version of the sort of system we saw in Czech; we also use this example to explain some conventions of representation. In this system the four classes are labelled A–D, and there are four morphosyntactic values, labelled

Table 6.3. *Grid system*

a. four-value system						b. distilled			
	A	B	C	D		A	B	C	D
1	a	b	c	d	→	a	b	c	d
2	e	f	g	h					
3	i	j	k	l					
4	m	n	o	p					

Table 6.4. *Hierarchical system*

	A	B	C	D
1	a	b	c	d
2	e	e	f	f

Table 6.5. *Cross-classifying system*

	A	B	C	D
1	a	a	b	b
2	e	f	f	e

1–4. Each class is distinct from all the others throughout the entire paradigm, which we have provisionally represented in Table 6.3 by a distinct letter in each cell. (Each letter represents a fully inflected form.) We call this a *grid* system. For each morphosyntactic value we have a contrast of four distinct forms, corresponding to the four inflection classes. Therefore, we can observe the structure of the inflection class system by looking at just one value; this view corresponds to what Stump & Finkel (2013) call a *distillation*.

Clearly, so long as one value within a system is configured as in Table 6.3b, the status of the system as having at least four inflection classes is ensured, even if these distinctions are conflated elsewhere. In Table 6.4, which we call a *hierarchical* system, although value 1 is realized by four different forms (*a*, *b*, *c*, *d*), value 2 has only two distinct forms (*e*, *f*). We call this hierarchical because the classes defined over the forms realizing one value (value 2) subsume the classes defined over the forms realizing the other value (value 1).

All other systems result from cross-classification. In Table 6.5, each value has only two distinct forms, but form *a* for value 1 can co-occur in a paradigm with either form *e* or *f* for value 2, and form *b* for value 1 can likewise co-occur with either form *e* or *f* for value 2. There are thus more classes than there are forms for any value.

Within these last two systems there is still room for further structural elaboration, even within the confines of our mini-system of four inflection classes. The hierarchical system can accommodate one more level, which subsumes all four classes (Table 6.6). We might call this system *maximal*, in that it exploits this type of organization to the maximum. In a similar fashion, the cross-classifying system in Table 6.5 can be extended. That simple system involved the class grouping {AB}, {CD} for value 1, and {AD}, {BC} for value 2. But a third exclusive grouping of these four classes is also possible – namely, {AC}, {BD}. If we add a third value to the system with this pattern, we arrive at the maximal cross-classifying system in Table 6.7.

We address how these types fare against explicit complexity measures in Chapter 7, but in the purely graphic descriptive terms employed here we have an increasing scale of complexity. The grid system can be represented in a single row (the class labels are just annotation), whereas the hierarchical system requires two rows, to account for the two distinct configurations. In order to adequately represent the simplest cross-classifying system – that is, to represent each distinct form in a single cell – would require a representation that wrapped around so as to bring classes D and A adjacent to each other, say by turning it into a squat Rubik's cube. The maximal cross-classifying system in

Table 6.6. *Maximal hierarchical system*

	A	B	C	D
1	a	b	c	d
2	e	e	f	f
3	g	g	g	g

Table 6.7. *Maximal cross-classifying system*

	A	B	C	D
1	a	a	b	b
2	e	f	f	e
3	g	h	g	h

Table 6.7 does not lend itself to any adequate representation in two or three dimensions, not in these terms at any rate. (Our point is not that it would necessarily be impossible to represent, simply that, given a fixed set of representational conventions, it becomes more unwieldy.)

We are unlikely to find many actual inflectional systems that instantiate these idealized canonical types, any more than perfect examples of the traditional typological categories of agglutinative or fusional languages. But they are a useful way of modelling the elements that compose heterogeneous inflectional systems. They emerge in pure form from when we subdivide paradigms along either of their constituent dimensions, – that is, in terms of morphosyntactic values, or in terms of lexical sets.

As an example of morphosyntactic subdivision, let us decompose the constituents of the system of subject prefixes in Santa Ana Keres verbs (briefly discussed in section 4.4). Table 6.8 gives fourteen inflection classes spread out over three mood values (indicative, dubitative, and hortative), three persons and three numbers.[1] In the first instance we can isolate several pairs of values that between them constitute grid systems with mutually interpredictable forms. These mostly involve dubitative forms and their corresponding indicative forms. These mini-grid systems are shown by boxes in Table 6.8, and separated out in Table 6.9, showing how the allomorphy within each pair lines up perfectly.

Next, some of the values constitute a hierarchical system between themselves, as shown in Table 6.10: starting with the first person dual indicative, the other values involve further conflation of the lexical classes. Most of the rest of the system then involves cross-classification of some sort; one pair of values will suffice to show how this works. Table 6.11 shows how three of the allomorphs found in the indicative first singular cross-classify with two of the allomorphs found in the first plural. Thus, the Santa Ana Keres system may look very different depending on which subset of values one chooses to look at.

The other way systems can be subdivided is along the lexical dimension. Consider the singular paradigms of the three Latvian nouns in Table 6.12. The word 'boy' patterns with 'sister' in the nominative and genitive, but with 'father' in the dative (the accusative and locative are identical for all three). The traditional account is that there are two inflection classes here: the class

[1] This represents a somewhat reduced set on two counts. First, Davis (1964) discusses other inflection classes, but does not include the full paradigm for all three numbers, so these have been omitted. Second, there are additional negative indicative forms, distinct only for first person. For the sake of (relative) simplicity these have been omitted.

Table 6.8. *Subject prefix classes in Santa Ana Keres*

	I	II	III	IV	V	VI	VII	VIII	IX	X	XI	XII	XIII	XIV
IND 1SG	cˀi	cˀi	s	s	s	sˀ	sˀ	səd	si	si	si	si	si	si
DUB 1SG	tˀi	tˀi	t	t	t	taʔ	taʔ	təd	ti	ti	ti	ti	ti	ti
HORT 1SG	kˀa	kˀa	k	k	k	kaʔ	kaʔ	nəd	ka	ka	ka	ka	ka	k
IND 2SG	š	š	gəẓ	ṣ	ṣ	gəẓ	ṣˀ	gəd	š	š	ṣ	ṣ	ṣ	ṣ
DUB 2SG	c	c	dʸəẓ	c̣	c̣	dʸəẓ	c̣aʔ	dʸəd	c	c	c̣	c̣	c̣	C̣
HORT 2SG	pi	pi	ẓ	c̣ˀ	c̣ˀ	ẓ	c̣ˀ	d	pi	ʔ	ʔ	ʔ	p	c̣ˀ
IND 3SG	z	z	kˀ	k	k	ga	kaʔ	z	c	z	g	g	g	k
DUB 3SG	d	d	čˀ	č	č	dʸa	čaʔ	d	t	d	dʸ	dʸ	dʸ	č
HORT 3SG	piʔ	piʔ	pˀ	p	p	pa	paʔ	p	piʔ	piʔ	piʔ	piʔ	piʔ	p
IND 1DU	səẓ	siʔ	səẓ	s	səẓ	səẓ	sˀ	səd	sˀi	səẓ	səẓ	səẓ	sˀi	s
DUB 1DU	təẓ	tiʔ	təẓ	t	təẓ	təẓ	taʔ	təd	tiʔ	təẓ	təẓ	təẓ	tiʔ	t
HORT 1DU	š	stiʔ	š	š	š	š	šˀ	st	stˀi	š	š	š	stˀi	š
IND 2DU	gəẓ	š	gəẓ	ṣ	gəẓ	gəẓ	ṣˀ	gəd	š	gəẓ	gəẓ	gəẓ	ṣ	ṣ
DUB 2DU	dʸəẓ	c	dʸəẓ	c̣	dʸəẓ	dʸəẓ	c̣aʔ	dʸəd	c	dʸəẓ	dʸəẓ	dʸəẓ	c̣	c̣
HORT 2DU	ẓ	pi	ẓ	c̣ˀ	ẓ	ẓ	c̣ˀ	d	pi	ẓ	ẓ	ẓ	p	c̣ˀ
IND 3DU	g	z	g	k	g	g	kaʔ	z	c	g	g	g	g	k
DUB 3DU	dʸ	d	dʸ	č	dʸ	dʸ	čaʔ	d	t	dʸ	dʸ	dʸ	dʸ	č
HORT 3DU	p	piʔ	p	p	p	p	paʔ	p	piʔ	p	p	p	piʔ	p
IND 1PL	səẓ	siʔ	səẓ	s	səẓ	səẓ	s	səd	sˀi	səd	səd	səẓ	sˀi	s
DUB 1PL	təẓ	tiʔ	təẓ	t	təẓ	təẓ	t	təd	tiʔ	təd	təd	təẓ	tiʔ	t
HORT 1PL	nəẓ	niʔ	nəẓ	n	nəẓ	nəẓ	n	nəd	nˀi	nəd	nəd	nəẓ	nˀi	n
IND 2PL	gəẓ	š	gəẓ	ṣ	gəẓ	gəẓ	ṣ	gəd	š	gəd	gəd	gəẓ	ṣ	ṣ
DUB 2PL	dʸəẓ	c	dʸəẓ	c̣	dʸəẓ	dʸəẓ	c̣	dʸəd	c	dʸəd	dʸəd	dʸəẓ	c̣	c̣
HORT 2PL	ẓ	pi	ẓ	c̣ˀ	ẓ	ẓ	c̣ˀ	d	pi	d	d	ẓ	p	c̣ˀ
IND 3PL	g	z	g	k	g	g	k	z	c	z	z	g	g	k
DUB 3PL	dʸ	d	dʸ	č	dʸ	dʸ	č	d	t	d	d	dʸ	dʸ	č
HORT 3PL	p	piʔ	p	p	p	p	p	p	piʔ	p	p	p	piʔ	p

Note: mutually interpredictable values are shown within boxes.
Source: Davis 1964.

Table 6.9. *Constituent grid systems within the subject prefix classes of Santa Ana Keres*

IND 1SG	cˀi	s	sˀ	səd	si		
DUB 1SG	tˀi	t	taʔ	təd	ti		
IND 2SG	š	gəẓ	ṣˀ	gəd	ṣ		
DUB 2SG	c	dʸəẓ	c̣aʔ	dʸəd	c̣		
IND 3SG	z	kˀ	ga	kaʔ	c	g	k
DUB 3SG	d	čˀ	dʸa	čaʔ	t	dʸ	č
IND 2DU	gəẓ	š	ṣ	ṣˀ	gəd		
DUB 2DU	dʸəẓ	c	c̣	c̣aʔ	dʸəd		
IND 3DU	g	z	k	kaʔ	c		
DUB 3DU	dʸ	d	č	čaʔ	t		
DUB 1PL	təẓ	tiʔ	t	təd	tiʔ		
HORT 1PL	nəẓ	niʔ	n	nəd	nˀi		
IND 2PL	gəẓ	š	ṣ	gəd			
DUB 2PL	dʸəẓ	c	c̣	dʸəd			
IND 3PL	g	z	k	c			
DUB 3PL	dʸ	d	č	t			

Table 6.10. *Constituent hierarchical system within the subject prefix classes of Santa Ana Keres*

IND 1DU	siʔ	sˀi	s	səẓ	səd	sˀ
DUB 1DU	tiʔ	tiʔ	t	təẓ	təd	taʔ
HORT 1DU	stiʔ	stˀi	š	š	st	šˀ
HORT 3DU	piʔ	piʔ	p	p	p	paʔ
HORT 3PL	piʔ	piʔ	p	p	p	p

Table 6.11. *Constituent cross-classifying system within the subject prefix classes of Santa Ana Keres*

IND 1SG	s	s	s	sˀ	sˀ	si	si
IND 1PL	s	səẓ	səẓ	səẓ	s	s	səẓ

Table 6.12. *Nouns of inflection classes I and II in Latvian*

	'father'	'boy'	'sister'
NOM SG	tēv-s	puik-a	mās-a
ACC SG	tēv-u	puik-u	mās-u
GEN SG	tēv-a	puik-as	mās-as
DAT SG	tēv-am	puik-am	mās-ai
LOC SG	tēv-ā	puik-ā	mās-ā

Table 6.13. *The three classes in Table 6.12, broken down into two sets of hierarchical classes*

	male humans		other nouns	
	class I	class II	class I	class II
NOM SG	-s	-a	-s	-a
GEN SG	-a	-as	-a	-as
DAT SG	-am		-am	-ai
ACC SG	-u		-u	
LOC SG	-ā		-ā	

represented by 'father' belongs to the masculine gender and the one represented by 'sister' is feminine. Nouns like 'boy' belong to the 'sister' class but denote male humans, and belong to the masculine gender, and take their dative singular form from the typically masculine paradigm.

If we distinguish two lexical sets here, consisting of nouns denoting male humans and everything else, we could get by with just two inflection class distinctions, as in Table 6.13. Within each lexical class, the inflection class systems are of the hierarchical type (Wurzel 1984/1989, Dressler et al. 2006). An alternative is to have two inflection classes (class I, for nouns like *tēvs* 'father', and class II for nouns like *māsa* 'sister'), with a condition that nouns denoting male humans take the dative singular in *-am*; we discussed conditions of this type, with more complex examples, in section 5.5.3.

6.2 Distributional Classes

Inflection classes characterized by the variable distribution of exponents bring with them a distinct set of questions, both in terms of the way the paradigm structure is characterized, and in terms of the treatment of morphosyntactic features.

In purely structural terms, the same types outlined in Tables 6.3–6.7 can be reproduced, albeit with a more limited scope, given that distributional variation is necessarily the interplay between two exponents (or one overt exponent versus zero, which amounts to the same thing). A limited grid system is possible given complementary distribution of exponents as in Table 6.14; note that, to show that this is a distributionally generated system, we need to show two lines, but a parsimonious representation of the classes could still get by with just one 'distillation'. A hierarchical system results from the extension of the distribution of one exponent from one to two cells (Table 6.15). Cross-classifying systems result from a combination of these two configurations (Tables 6.16 and 6.17).

Table 6.14. *Distributional classes in a grid system*

	A	B
1	a	b
2	b	a

Table 6.15. *Distributional classes in a hierarchical system*

	A	B
1	a	b
2	a	a

Table 6.16. *Distributional classes in a cross-classifying system*

	A	B	C	D
1	a	a	b	b
2	a	b	b	a

Table 6.17. *Distributional classes in a maximal cross-classifying system*

	A	B	C	D
1	a	a	b	b
2	a	b	b	a
3	a	b	a	b

Although they can generate the same sorts of paradigmatic structures, allomorphic and distributional classes differ from each other in the terms needed to characterize them. One obvious way to see this is through the principal parts needed to identify the classes. Consider the cross-classifying systems in Table 6.5 (allomorphic) and Table 6.16 (distributional). In both types, we need to list two principal parts. In the allomorphic type, it is enough to list the exponents as unordered sets: {ae}, {af}, {bf}, and {be}. In the distributional type, that is not enough: {aa} and {bb} are unequivocal identifiers of their class, but {ab} does not distinguish between class B and class D. These need to be annotated with information about what cells they appear in, along the lines {{a,1}{b,2}} and {{a,2}{b,1}} – that is, class B is the one where exponent a is used for value 1 and exponent b for value 2, and class D is the one where exponent a is used for value 2 and exponent b is used for value 1. Obviously, in actual language use, inflected words do not usually float around without any external indication of their morphosyntactic value, so it could be argued that this information is present in any case. But our purpose here is to explore properties of the system, and the fact is that, all else being equal, the distributional type requires more explicit information to describe it.

6.2.1 Feature Values

Distributional classes bring with them an important additional analytical question. That is, any kind of inflection class system, be it of the allomorphic or distributional type, involves the structure of the lexicon and its division into classes of words. Distributional classes additionally involve the structure of the feature system and its division into classes of values. This brings into focus the question of motivatedness versus arbitrariness in morphological structure; this is because some of these configurations present at least prima facie difficulties to analyses in which paradigm structure is motivated by direct reference to the values of morphosyntactic features, as opposed to being morphologically stipulated. We can begin by looking at two basic configurations, considering them in the light of possible 'motivated' analyses. In Table 6.18(a), exponent 'a' is restricted to value 1 in class A, whereas in class B it is found for both value 1 and value 2. Depending on which class we take as the starting point, we can construe the morphosyntactic value of a as being restricted (B → A) – for example, by under-specification and blocking, or expanded (A → B), as with impoverishment, which would erase the distinction between feature values 1 and 2. In Table 6.18(b), the inventory of exponents is the same, but their distribution in the two classes is complementary. A motivated interpretation is possible in terms of abstract feature values with some kind of toggle function (e.g., polarity, exchange rules).

All other types can be derived from these basic configurations, but they do raise new issues. Thus, Table 6.19 is an extension of Table 6.18(a) in which both exponents display a variable range of distribution. This makes a motivated interpretation more difficult, if not impossible: if both a and b can range over values 1 and 2 (classes B and C), how can they negotiate their coexistence in class A? Table 6.20 is a combination of the two systems in Table 6.18. It is also unfriendly to a motivated interpretation, due to the variety of relationships that have to be accommodated.

In practice, we find evidence of both motivated and arbitrary distributional classes. Consider first one extreme: in the paradigms in Table 6.21 from Eastern Armenian, a member of the Indo-European family, the suffix *-i* is found with the dative singular with 'mountain' and with both the dative and accusative singular with 'pupil'. This is clearly a motivated pattern: the 'pupil' pattern is characteristic of human nouns, and represents a more general pattern of

Table 6.18. *Distributional classes: basic systems compared*

a. hierarchical			b. grid		
	A	B		A	B
1	a	a	1	a	b
2	b	a	2	b	a

Table 6.19. *Distributional classes: expanded hierarchical system*

	A	B	C
1	a	a	b
2	b	a	b

Table 6.20. *Distributional classes: hierarchical and grid systems combined*

	A	B	C
1	a	b	a
2	b	a	a

Table 6.21. *Eastern Armenian inanimate ~ animate contrast*

	'mountain'	'pupil'
NOM SG	sar	ašakert
ACC SG	sar	ašakert-i
DAT SG	sar-i	ašakert-i
ABL SG	sar-ic'	ašakert-ic'
INS SG	sar-ov	ašakert-ov

Source: Dum-Tragut 2009.

Table 6.22. *Ingush nouns*

	'child'	'friend'
ERG PL	bier-azh	dottagh-azh
ABS PL	bier-azh	dottagh-ii
GEN PL	bier-ii	dottagh-ii
DAT PL	bier-azh-ta	dottagh-azh-ta
ALL PL	bier-azh-ka	dottagh-azhka
INS PL	bier-azh-ca	dottagh-azhca
LAT PL	bier-egh	dottagh-egh
CSN PL	bier-el	dottagh-el

Source: Nichols 2011.

differential object marking in the language, in which the direct object case for any human noun is identical to the dative, whatever its form, whereas the direct object case for any non-human noun is the same as its nominative.

The Armenian pattern is one that, arguably, should be construed as a question of morphosyntax, and not of morphology as such. But the division is not always clear. Russian object marking, for example (see section 5.3), bears a strong resemblance to Armenian, but the system is intertwined with semantically arbitrary morphological classes. Still, other examples mimic the surface patterns without sharing the same sort of motivation in any synchronically recoverable way. For example, in the plurals of most nouns in the Nakh language Ingush (Table 6.22), the absolutive (nominative), ergative, and genitive share between them two suffixes. The ergative takes *-azh* and the genitive *-ii*, but the suffix of the absolutive varies: some nouns take *-azh*, others take *-ii*. In one sense it resembles the pattern seen in Armenian, in that the two patterns differ in the presence or absence of a distinction between the cases that mark core arguments. But the contrast has no plausible synchronic basis in differential argument marking: the semantic membership of the classes is arbitrary,

Table 6.23. *Otomí varieties compared*

	a. Acazulco Otomí		b. Tilapa Otomí	
	class II	class III	class II	class III
REAL INCPL 1	drá	drádí	trá	trátí
REAL INCPL 2	grá	grádí	grá	grátí
REAL INCPL 3	ra	radí	ra	rati
REAL CPL 1	dídí	dí	túdú	tú
REAL CPL 2	gídí	gí	gúgú	gú
REAL CPL 3	bi	bi	bi	bi
IRR CPL 1	gidí	gidi	gutu	giti
IRR CPL 2	gidí	gidi	gugu	giti
IRR CPL 3	di	di	ti	ti

Source: Hernández-Green 2015, Palancar 2012.

and besides, this contrast only affects the plural: all three cases are distinct in the singular.

At the other extreme we find distributional classes that have no plausible basis in morphosyntactic alternations. Consider the suffix *-di* in the Acazulco Otomí class II and class III inflectional formatives (see section 2.1.3 for further discussion of the paradigm). Table 6.23(a) shows a fragment of the paradigm that illustrates the range of patterns. The two classes differ solely in the distribution of *-di*. The paradigm-internal pattern of distribution does lend itself to interpretation in featural terms: it is used for first and second person in the irrealis completive of both classes, and in the realis completive just of class II, whereas in the realis incompletive it is used for all three persons in class III. Nor do the classes themselves constitute coherent semantic or syntactic families. The variant patterns thus appear to be completely unmotivated. Comparative evidence gives further support to a morphological interpretation. Tilapa Otomí in Table 6.23(b) is a closely related but more conservative variety. The corresponding inflectional formatives have suffixes in exactly the same cells as in Acazulco Otomí, but the forms are different between the two classes. Not only are the forms different, but the class II suffix further distinguish person: first person *-du* or *-tu*, and second person *-gu*. Thus, in Tilapa, suffixation in the two classes operates on two distinct principles: class II has person-marking suffixes used with certain aspect-mood values, whereas class III has an invariant suffix *-ti* found in a distinct but overlapping set of aspect-mood values.[2] What appears

[2] It is tempting here to say that *ti* is used for all three persons, and that its apparent restriction to first and second person in the irrealis completive is due to morphological haplology, on the

Table 6.24. *Hypothetical intermediate system between Tilapa and Acazulco Otomí: t → d, u → i*

			class II	class III
REALIS	INCPL	1		di
REALIS	INCPL	2		di
REALIS	INCPL	3		di
REALIS	CPL	1	di	
REALIS	CPL	2	* gi	
REALIS	CPL	3		
IRR	CPL	1	di	di
IRR	CPL	2	* gi	di
IRR	CPL	3		

to have happened in Acazulco was the merger of these two distinct sets of suffixes into a single suffix with two distinct distributions.

We reconstruct this as the result of analogical change prompted by phonological change. Observe that in the formative base there are two key differences between Tilapa and Acazulco: (1) Tilapa /t/ is realized as Acazulco /d/ and (2) Tilapa /u/ is realized as Acazulco /i/. Transposing these two changes to the suffixes yields the paradigm in Table 6.24. Full absorption of the class II suffix series would require only analogical change of the second person suffixes to correspond to the single *-di* suffix found everywhere else (the proportional analogy would be: class III first person *di*: class III second person *di*:: class II first person *di*-: x). If this scenario is correct, then the Acazulco classes represent two wholly unrelated distributions that are realized through what has been construed as the same suffix.

Defined in conventional formal terms, the distinction between motivated and arbitrary is an all-or-nothing affair. But matters may be rather more subtle than that, as shown by some examples where the alternation between classes is constrained by feature values without being directly reducible to the manipulation of feature structure. Consider first the Nuer plural suffix described in section 4.2. It occurs in two paradigmatic configurations, either as an oblique plural case suffix, or simply as a plural suffix. Even if the nouns involved fell into identifiable semantic classes (which they do not), there is no plausible scenario that would account for the presence or absence of case distinctions

assumption that the base formative in the third person is *ti* (as in class II), which is then suffixed with *ti*, whereby **titi* → *ti*. If this interpretation has any synchronic or diachronic validity, it is also true that it no longer applies to Acazulco (i.e., first person realis incompletive *dídí*).

Table 6.25. *Distributional classes in Seri; third person dependent realis forms*

		'go in zigzag'	'unfasten'	'glue'	'be soft'
SG SBJ	SG EVENT	kʷtmonxk	kʷitá:fp	itaksípχ-a	tíχpχ-ax
SG SBJ	PL EVENT	kʷtmónax-ox	kʷitá:fip-ox	itaksípχ-ox	tíχpχ-ołka
PL SBJ	SG EVENT	kʷtmónx-ox	kʷitató:fip-ox	itaksípχ-ołka	tíχipχ-ołka
PL SBJ	PL EVENT	kʷtmónx-ox	kʷitató:fip-ołka	itaksípχ-ołka	tíχipχ-ołka

Source: Moser & Marlett 2010, Marlett 2015.

on semantic or functional grounds. Worse, stem alternations may mark case independently, so the absence of suffixal case marking is not the equivalent of the absence of case marking. This suggests we should view the alternation in the same light as that of Acazulco Otomí, as two morphologically stipulated patterns. Intuitively, though, that seems wrong. There is a clear thread running through both distributions – the suffix is restricted to the plural – and by the normal criteria we use to talk about morphological forms it is some kind of plural marker. What must additionally be stipulated is whether it also functions as a case marker, so that the morphological arbitrariness is nested within a morphosyntactically motivated pattern (see Corbett 2016).

In Seri, distributional classes appear to result from the variable calibration of the mapping between morphosyntax and morphology. A representative set of paradigms is given in Table 6.25, distinguishing subject number and event number, as discussed in section 3.2. The paradigm involves both suffixation and, in some cases, stem alternations; our focus here is on the suffixes. All the paradigms here show the use of either or both of the suffixes *-tox* and *-tołka* (recall from section 6.1 that suffix-initial /t/ is deleted under phonologically predictable conditions); these mark plural values – that is, any cell in the paradigm marking plural event or plural subject. Clearly, we cannot attribute any fixed value to either of the suffixes, since both *-tox* and *-tołka* can be used on their own for all plural values, and they can also co-occur within the same paradigm. But crucially, where they do co-occur, they follow a strict hierarchy: given a morphosyntactic hierarchy [SG SBJ, PL EVENT] < [PL SBJ, SG EVENT] < [PL SBJ, SG EVENT], *-tox* always occurs to the left of *-tołka*. Both the forms and the morphosyntactic values give a clue as to what might be going on. Historically, the suffix *-tołka* is probably composed of *-tox* plus another plural suffix *-ka* (compare the singular subject, singular event form *-i:mx* 'be clear' with the corresponding plural subject form *-i:mł-ka*, with the morphophonologically regular alternation of /x/ ~ / ɬ/ before /k/). This suggests

Table 6.26. *Summary of plural suffix patterns in Table 6.25*

		'go in zigzag'	'unfasten'	'glue'	'be soft'
SG SBJ	SG EVENT				
SG SBJ	PL EVENT	tox	tox	tox	tołka
PL SBJ	SG EVENT	tox	tox	tołka	tołka
PL SBJ	PL EVENT	tox	tołka	tołka	tołka

that *-tołka* is in some sense a multiply marked plural – and thus, "more plural" than *-tox*. The associated morphosyntactic values also indicate that *-tołka* is more plural: in those instances in which its use is restricted to a single cell, as with 'unfasten', it is the maximally plural cell.

The implicit cline of degrees of plurality sketched in Table 6.26 is thus both morphological and morphosyntactic, but the two are independent of each other. Alternatively expressed, we can take the suffixes and arrange them into a paradigm, but the precise link to the morphosyntactic paradigm may remain undetermined. The distributional inflection classes are thus defined as different mappings between morphosyntactic and morphological plurality.

6.3 Mixed Patterns

We have looked so far at the paradigmatic structure of inflection classes along two dimensions as defined by allomorphic and distributional variation. Obviously, the two can be combined, with a corresponding increase in the number of classes and the complexity of the system. The aspect-mood suffixes of Gulmancema (whose tonal alternations were discussed in section 2.3) represent a near-perfect union of allomorphic and distributional classes. The paradigm of any one verb involves the presence or absence of a suffix *-ni*, *-di*, *-li* or *-gi*; plus some verbs are completely unsuffixed.[3] The suffixes can occur in any of three distributions: in the imperfective, in the perfective, or in the aorist and perfective. These patterns are apparently arbitrarily specified. The three distributional patterns co-occur with three of these suffixes (Table 6.27), whereas the fourth, *-gi*, is a dedicated imperfective suffix (Table 6.28).

[3] Following Ouoba (1982), we construe verb stems as either monosyllabic (e.g., *púá* 'hit.POT/PFV') or disyllabic (*bòbì* 'set out.POT'). Judging from the corpus of verbs given by Naba (1994), disyllabic verbs all end in *-i*, and the resulting /Ci/ is often superficially the same as one of the suffixes (*baalí* 'caress.POT/PFV'), but is clearly distinct (*baal-dí* 'caress.IPFV'); note that such deletion of stem-final *-i* (but not suffix-final *-i*) is also regular under certain syntactic conditions (Ouoba 1982).

Table 6.27. *Gulmancema allomorphic-cum-distributional classes*

	'tap on the head'	'return'	'do'	'pass'	'love'	'hear'	'fall'	'give birth'	'plant'
AOR[a]	tuà	goá	tié-ní	cié	bua	gbà-dì	bà	ma	bù-lì
IPFV	tua-ní	goâ	tie	cié-dí	buà	gbà	baà-lí	ma	bu
PFV	tuà	goá-ní	tié-nì	ciê	bua-dì	gba-dì	bà	ma-lì	bù-lí

[a] Naba (1994) defines the *aorist* as the potential mood, unmarked for aspect, opposed to the imperfective and perfective realis.
Source: Naba 1994, Ouoba 1982.

Table 6.28. *Gulmancema dedicated imperfective suffix*

	'lick'
AOR	liâ
IPFV	liá-gi
PFV	liá

Table 6.29. *Prefix and suffix allomorphy in Ayoreo*

	'shout'	'say'
1SG	yi-bi-te	yi-go
2SG	ba-bi-te	ba-go
3	ti-bi-te	chi-go
1PL	yi-bi-co	yi-go-go
2PL	uaca-bi-cho	uaca-go-yo

Source: Ciucci 2007.

Seri verbal inflection takes this one step further, due to the greater number of allomorphs (see section 3.2) and the relative freedom with which they can be combined in the paradigm (see section 6.2). Considering just the inflectional suffixes, the verbs represented in Moser and Marlett's (2010) dictionary fall into over 250 different inflection classes (Baerman 2016).

6.4 Combining Systems

Inflection can target multiple aspects of a word form, each of which is liable to fall into different inflection classes. Consider the verb paradigms in Table 6.29

Table 6.30. *Different prefixes with same suffixes in Ayoreo*

	'steal'[a]	'fear'
1SG	y-o/uria	yi-todo
2SG	b-oria	ba-todo
3	t-oria	todo
1PL	y-oria-go	yi-todo-go
2PL	uac-oria-yo	uaca-todo-yo

[a] Final vowel of prefix regularly deletes before vowel-initial stems.
Source: Ciucci 2007.

Table 6.31. *Different suffixes with same prefixes in Ayoreo*

	'dispatch'	'take away'[a]	'sit'	'knock down'
1SG	yi-ro-que	ñi-jno-que	y-aca-re	yi-guisa-re
2SG	ba-ro-que	ma-jno-que	b-aca-re	ba-guisa-re
3	chi-ro-que	chi-jno-que	ch-aca-re	chi-guisa-re
1PL	yi-ro-co	ñi-jno-jo	y-aca-go	yi-guisa-co
2PL	uaca-ro-cho	uaca-jno-cho	uac-aca-yo	uaca-guisa-yo

[a] Prefixes here show phonologically regular nasal harmony conditioned by the stem.
Source: Ciucci 2007.

from Ayoreo, a Zamucoan language of Paraguay and Bolivia. Subject person-number is marked both by prefixes and suffixes. In the verbs shown here, the prefixes differ (third person *ti-* vs. *chi-*) as well as the suffixes (*-te*/*-co*/*-cho* vs. Ø/*-go*/*-yo*). Judging by the examples in Ciucci (2007), prefix allomorphy and suffix allomorphy cross-classify freely. For example, the same suffix pattern as with 'say' also occurs with the third-person prefixes *ti-* and Ø, as in Table 6.30, and the same prefix pattern occurs with various suffix patterns (Table 6.31).

In terms of the typology outlined in section 6.1, this is a cross-classifying system, though the domain is obviously different. In this case, the reasons are fairly clear: the prefixes and suffixes are not adjacent and thus have no phonological influence on each other, and the principles originally governing their allomorphy were unrelated. The suffix classes go back historically to derivational suffixes (Ciucci & Bertinetto 2015: 35f), which merged phonologically with the plural subject suffixes. Prefix allomorphy operated on different, unrelated principles (not recoverable), and did not interact with the termination of the stem, or the suffixes.

Table 6.32. *Estonian strengthening stem alternation (strong stem shaded); repeated from Table 2.13*

	'thought'		'tooth'		'heart'	
	SG	PL	SG	PL	SG	PL
NOM	mõte	\`mõtte-d	hammas	hamba-d	süda	südame-d
PART	mõte-t	\`mõtte-id	hammas-t	hamba-id	süda-t	südame-id
GEN	\`mõtte	mõte-te	hamba	hammas-te	südame	süda-te

Table 6.33. *Estonian weakening stem alternation (strong stem shaded); repeated from Table 2.14*

	'bag'		'war'	
	SG	PL	SG	PL
NOM	\`kott	koti-d	sõda	sõja-d
PART	\`kotti	\`kotte	sõda	sõda-sid
GEN	koti	\`kotti-de	sõja	sõda-de

Conversely, distinct inflectional subsystems can stand in an implicative relationship to each other. This is the case with the stem alternation and suffixation patterns in Estonian, alluded to in sections 2.1.1 and 2.1.3. A substantial portion of the Estonian nominal lexicon follows either of two stem alternation patterns, the strengthening type (Table 6.32) and the weakening type (Table 6.33); other nouns lack a stem alternation, and there are other, minor patterns. The two-stem alternation classes are likewise associated with distinct patterns of case-number suffixation: (1) the partitive singular is *-t* with strengthening stems, zero with weakening stems; (2) the partitive plural is *-id* with strengthening stems, whereas, with weakening stems, either a stem-final vowel alternation or the suffix *-sid* is found, their distribution largely depending on prosodic properties of the stem (doublet forms are not infrequent, so that, for example, *kotti-sid* and *sõdu* are also found; Erelt, Erelt, and Ross 1997: types M140, M144); and (3) the genitive plural is *-te* with strengthening stems, *-de* with weakening stems.

For the most part, the allomorphy of the suffixes was conditioned by the sort of phonological developments outlined in section 2.1, though the *-sid* allomorph of the partitive plural also involved co-option of a stem-final /s/ found in certain nominals (Rätsep 1979: 21). But as things stand now, there is no direct predictability between the phonological properties (in terms of

Table 6.34. *Estonian suffix class opposition (same as in Tables 6.32 and 6.33, but with non-alternating stems)*

	'pupil (school)'		'tail'	
	SG	PL	SG	PL
NOM	õpik	õpiku-d	saba	saba-d
PART	õpiku-t	õpiku-id	saba	saba-sid (*or* sabu)
GEN	õpiku	õpiku-te	saba	saba-de

Table 6.35. *Paradigm of 'search'*

	PRS	CPL	FUT
3	$hnangh^{2}$	ma^{2}-$hnangh^{2}$	$hnangh^{1}$
1SG	$hnangh^{12}$	ma^{2}-$hnangh^{1}$	$hnangh^{13}$
1PL	$hniangh^{2}$	ma^{2}-hniángh^{3}	hniángh^{3}
2	hnángh^{2}	ma^{2}-$hniangh^{2}$	hnángh^{3}

Source: Merrifield & Anderson 2007: 697.

segmental and prosodic structure) of a particular stem form and the suffix allomorph it takes. Thus, partitive singular *süda* and *sõda-t* are both CVCV stems with a short initial syllable, but take different suffixes (and similarly with genitive plural *süda-te* vs. *sõda-de*), and partitive plural *`mõtte-id* and *`kotte (or `kotti-sid)* are both CVCV stems with a geminate medial consonant and overlong initial syllable, but take different suffixes. The direction of the condition is between the stem alternation PATTERNS and suffixation, and not the individual stem forms (compare section 5.5.9). Note that the implication is unidirectional, because the same suffix patterns also occur with non-alternating stems (Table 6.34).

The Ayoreo and Estonian examples represent two extremes: complete lack of interaction, and a categorical implication. But the usual case lies somewhere in between. A particularly interesting instance of this comes from San Pedro Tlatepuzco Chinantec – closely related to Lealao Chinantec, discussed in section 4.4. (All examples taken from Merrifield & Anderson 2007: 691–99 unless otherwise noted.). In section 7.3 we present an implemented model of its rich system of tone alternations. But what interests us here is the relationship between the different types of exponence. A representative paradigm is given in Table 6.35. TAM and subject person-number inflection involves several intersecting processes (1) tone/stress alternations, (2) a segmental stem alternation

Table 6.36. *Examples of class A verbs in Tlatepuzco Chinantec*

	'shave'			'tie up'			'cough'		
	PRS	CPL	FUT	PRS	CPL	FUT	PRS	CPL	FUT
3	ta^{12}	ta^{1}		ʔngieng^{12}	ʔngiéng^{2}	ʔngieng^{1}			
1SG			ta^{13}		ʔngieng^{1}	ʔngieng^{13}		húʔ^{2}	
1PL	ta^{2}		tá^{3}			ʔngiéng^{3}			húʔ^{3}
2		ta^{3}		ʔngieng^{2}					

(*hnagh* versus *hniagh*) and (3) prefixation.[4] The first two processes – tone/stress and stem alternations – fall into a number of different classes, which display some interesting interactions; prefixation is consistent across all verbs, and will be ignored in the following discussion.

The tone/stress paradigm involves various level and contour tones (this is represented by superscript numerals, with 1 being the lowest and 3 the highest), plus a prosodic feature termed 'ballistic stress' in much of the literature on the Chinantecan languages, represented here with an acute accent. All told, there are something over 80 tone/stress types, if one considers both the pattern of alternation and the actual tone/stress values. But Merrifield & Anderson suggest that these can be classed into larger macro-classes, and it is at this level that the noticeable interactions with stem alternations take place. These macro-classes are *distributional* macro-classes, based on the way the person-number paradigm is partitioned. Class A is defined as having, somewhere in the paradigm, a distinction between first and second person. (The rather unconventional order of person values in the following tables has been adopted to maximize the adjacency of identical forms.) This can take many shapes (Table 6.36). For example, 'shave' has five distinct forms with a complex paradigmatic distribution, 'tie' has four forms with a very different distribution, and 'cough' has just two forms, but, crucially for this classification, the second person and the first-person plural are distinguished in the completive.

Class B distinguishes the third person from the rest in at least some portion of the paradigm, but first and second person are always identical (Table 6.37). For example, with 'dry' there are just two forms, one for third person and one for the rest, whereas with 'take', the third-person form distinguishes between the present and the rest.

[4] In addition, there is suffixation of a glottal stop in the second person. As far as can be told from the information in Merrifield & Anderson (2007), second-person suffixation applies to all verbs whose stems do not already end in a glottal stop and so does not concern us here.

Table 6.37. *Examples of class B verbs in Tlatepuzco Chinantec*

<table>
<tr><th></th><th colspan="3">'dry'</th><th colspan="3">'take'</th></tr>
<tr><th></th><th>PRS</th><th>CPL</th><th>FUT</th><th>PRS</th><th>CPL</th><th>FUT</th></tr>
<tr><td>3</td><td colspan="3">tsén^{2}</td><td>ʔenʔ^{12}</td><td colspan="2">ʔenʔ^{1}</td></tr>
<tr><td>1SG</td><td colspan="3" rowspan="3">tsen^{3}</td><td rowspan="3"></td><td colspan="2" rowspan="3">ʔenʔ^{13}</td></tr>
<tr><td>1PL</td></tr>
<tr><td>2</td></tr>
</table>

Table 6.38. *Examples of class C verbs in Tlatepuzco Chinantec*

<table>
<tr><th></th><th colspan="3">'disappear'</th><th colspan="3">'reappear'</th><th colspan="3">'revive'</th></tr>
<tr><th></th><th>PRS</th><th>CPL</th><th>FUT</th><th>PRS</th><th>CPL</th><th>FUT</th><th>PRS</th><th>CPL</th><th>FUT</th></tr>
<tr><td>3</td><td rowspan="4">kionʔ^{12}</td><td rowspan="4">kionʔ^{1}</td><td rowspan="4">kionʔ^{13}</td><td rowspan="4">hniang^{2}</td><td colspan="2" rowspan="4">hniang^{1}</td><td colspan="2" rowspan="4">hiog^{2}</td><td rowspan="4">hiog^{1}</td></tr>
<tr><td>1SG</td></tr>
<tr><td>1PL</td></tr>
<tr><td>2</td></tr>
</table>

Class C has no distinction of person, only of tense aspect (Table 6.38). This may be the maximal three-way distinction, as in 'disappear', or it may involve conflation, as in 'reappear', where completive and future are identical, or in 'revive', where present and completive are identical. (There are also verbs where all three are conflated.)

Stem alternations likewise occur in a large number of different patterns, and it will also be useful to abstract out some general trends. We divide the patterns described by Merrifield & Anderson (2007) into two broad types. Simple alternations as shown in Table 6.39 are those that involve only person values or only TAM values. The verb 'accompany', for instance, has a stem associated with third person (as well as a unique first singular future stem), whereas the verb 'tire' has dedicated stems for the first person singular and second person. The stem alternation in the verb 'release', on the other hand, distinguishes future from the other TAM values.

Compound alternations as in Table 6.40 are those that can only be described by invoking both person-number and TAM values; thus, first-person plural and second-person completive as in 'look for' or first-person plural and second-person present and future as in 'sell'. (There are still further types that involve combinations of these patterns.)

Table 6.39. *Simple stem alternations*

	'accompany'			'tire'		
	PRS	CPL	FUT	PRS	CPL	FUT
3	hmøi^{2}	hmøi^{2}	hmøi^{2}	hóh^{12}	hóh^{1}	hóh^{13}
1SG	hmái^{12}	hmái^{1}	hnangh^{13}	hóg^{12}	hóg^{1}	hóg^{13}
1PL	hmái^{12}	hmái^{13}	hmái^{13}	hóh^{12}	hóh^{1}	hóh^{13}
2	hmái^{12}	hmái^{1}	hmái^{13}	huanh^{12}	huanh^{1}	huanh^{13}

	'release'		
	PRS	CPL	FUT
3	tug^{12}	tug^{12}	tiug^{12}
1SG	tug^{12}	tug^{12}	tiug^{12}
1PL	tug^{12}	tug^{12}	tiug^{12}
2	tug^{12}	tug^{12}	tiug^{12}

Table 6.40. *Compound stem alternations*

	'look for'			'sell'		
	PRS	CPL	FUT	PRS	CPL	FUT
3	hnangh^{2}	hnangh^{2}	hnangh^{1}	hnai^{12}	hnai^{1}	hnai^{1}
1SG	hnangh^{12}	hnangh^{1}	hnangh^{13}	hnai^{12}	hnai^{1}	hnai^{13}
1PL	hniangh^{2}	hniángh^{3}	hniángh^{3}	hnøi^{2}	hnói^{3}	hnói^{3}
2	hnángh^{2}	hniangh^{2}	hnángh^{3}	hnøi^{2}	hnai^{3}	hnói^{3}

Given this division of tone/stress paradigms into A, B, and C and of stem alternations into simple and compound, a quite striking interaction can be seen, shown in Table 6.41. Only a minority of verbs display a stem alternation, but among those that do, the type of alternation is partly predicted by the tone/stress class. Class A favours compound stem alternations, class B is evenly divided between the two types, and class C almost exclusively takes the simple type of stem alternation.

There seems to be a relationship between the complexity of the tonal alternation and the complexity of the stem alternations. This would make sense if we could derive the stem alternations from the tone/stress alternations, because complexity in the former would have to be licensed by complexity in the latter. But

Table 6.41. *Distribution of lexemes in Merrifield & Anderson (2007)*

<table>
<tr><td></td><td></td><td colspan="2">stem alternation type</td><td rowspan="2">total size of class</td></tr>
<tr><td></td><td></td><td>compound</td><td>simple</td></tr>
<tr><td rowspan="3">tone/ stress</td><td>class A</td><td>86</td><td>10 (person only)</td><td>360</td></tr>
<tr><td>class B</td><td>14</td><td>14</td><td>251</td></tr>
<tr><td>class C</td><td>1</td><td>22</td><td>135</td></tr>
</table>

Note: The figures for 'total size of class' are those explicitly given by Merrifield & Anderson (2007) in the grammar section of their dictionary. The figures for the different stem-alternation types are derived from the dictionary entries themselves.

Table 6.42. *Orthogonal systems in a class C verb ('obtain')*

<table>
<tr><td></td><td colspan="3">tone</td><td></td><td colspan="3">stem alternation</td></tr>
<tr><td></td><td>PRS</td><td>CPL</td><td>FUT</td><td></td><td>PRS</td><td>CPL</td><td>FUT</td></tr>
<tr><td>3</td><td rowspan="4">2</td><td colspan="2" rowspan="4">1</td><td rowspan="4">+</td><td colspan="3">jniuh</td></tr>
<tr><td>1SG</td><td colspan="3" rowspan="3">jniang</td></tr>
<tr><td>1PL</td></tr>
<tr><td>2</td></tr>
</table>

there is no direct mapping from tone/stress to stem alternation. Consider class C. Verbs of this class only mark TAM with tone/stress, never person-number, and they only permit simple stem alternations. But the simple stem alternations they permit may well involve person alternations, as shown in Table 6.42. Thus both systems are simple, in a sense, but the simplicity is not congruous. Likewise, the actual patterns displayed by compound stem alternations do not mirror the distinctions made by the tone/stress classes that seemingly license them. Consider the verb in Table 6.43. Its tone/stress paradigm distinguishes person values and so permits a compound stem alternation. But the stem alternation it displays occurs entirely with a portion of the paradigm in which it is prosodically invariant. The distinctions made by stem alternations do not mirror those made by tone/stress. There is no known diachronic explanation for the relationship; there is an intriguing correlation of paradigmatic complexity between the two systems; however, there is no mechanism we are aware of that will account for it.

Table 6.43. *Orthogonal systems in a class B verb ('call')*

	PRS	CPL	FUT
3	2		1
1SG			
1PL		3	
2			

\+

stem alternation

	PRS	CPL	FUT
3		tai	
1SG			
1PL		tei	
2	tai		tai

6.5 Conclusion

On the face of it, the term *inflection class* conjures up images of a discrete object in opposition to other similarly discrete objects. But the coherence of classes typically breaks down when we look more closely at paradigms: the way the lexicon is divided into sets will vary from one morphosyntactic value to another, or from one morphological component to another. The paradigm of any one lexeme may then represent a network of shifting alliances. For its complexity and variety, we have tried here to isolate the key typological parameters along two dimensions: that of allomorphic versus distributional classes on the one hand, and the three-way distinction between grid, hierarchical, and cross-classifying arrangements on the other. Real-life inflectional systems are almost invariably some combination of these, so that the overall typological profile of a given language is composed of these smaller parts. How this plays out in an overall measure of inflectional complexity is a topic we take up in Chapter 7.

7 *Lexicon and Grammar*

In Chapter 6 we presented the three canonical types of paradigm structure: grid systems, hierarchical systems and cross-classifying systems. In grid systems each inflectional class is distinct from every other class throughout its paradigm. This means that we only really need to look at one morphosyntactic value to determine the inflection class divisions. In a hierarchical system the maximal set of class distinctions is found for one morphosyntactic value, while other values display a smaller number, subsuming class distinctions found in the maximal set. In a cross-classifying system, the lexicon is divided in different ways for each value; if we think in terms of allomorphy, this means that the number of classes needed to describe the whole system exceeds the number of allomorphs found for any single value. In this chapter we show how these canonical types fit within a three-dimensional typology of complexity that links grammar and the lexicon. The simple structure of the grid system provides the basis for a set of rules (the grammar), because the relationship between exponents and values is reliable: given knowledge of a value and a form it is possible to predict another value and its form, and this works in both directions (i.e., it is bi-directional). The cross-classifying system relies on listing (the lexicon), because the relationship between exponents and values is unreliable: knowledge of a value and a form is insufficient to predict another value and its form. The hierarchical system is a compromise between lexicon and grammar, because the relationship between exponents and values is largely, but not entirely, reliable: given knowledge of some values and forms it is possible to predict other values and their forms, but this works only in one direction (i.e., it is uni-directional).[1]

[1] How one captures these different properties in psycholinguistic models is an interesting question. Phonological similarity, as captured in Albright's (2002) minimal generalization learner (MGL) for instance, does not have to work in both directions. Similarly, uni-directionality is a property of systems involving defaults, because the canonical default is not predictive, but is highly predictable (Brown & Evans 2012:139–140). The question is how much reliance is placed on the intrinsic predictability of defaults, and how much on the implicative power of other forms to predict them. It is worth noting in this regard that Veríssimo & Clahsen (2014)

Having considered the relationship between the three different systems and different types of complexity in section 7.1, we show in section 7.2 that when we measure the different structures, this parallels our observation regarding the intermediate status of hierarchical systems in relation to grid systems and cross-classifying systems. Grid systems, because they can be underpinned by a grammar, are high in what we term 'organization'. Cross-classifying systems are high in the opposite type of complexity, which we term 'emergent complexity', because the relationship between values is not predictable. However, in another sense they are simple, because they rely *solely* on listing. Viewed from this perspective they are similar to grid systems, as these can rely *solely* on rules. Because they are neither one thing nor the other, hierarchical systems are high in 'central system complexity', which increases as the contributions of the grammar and lexical listing approach each other. Once we have illustrated the idealized types we present a composite case study in section 7.3 of the tonal classes in Tlatepuzco Chinantec, giving two complementary analyses: one is based on a structure intermediate between a single paradigm cell and the whole paradigm, the 'inflectional series'; the other is based on defaults and principal parts, but allowing for a reduction in the number of principal parts down to zero. An interesting aspect of this model is that even though a lexical item may be assigned to the wrong inflectional class, because of default assignment based on little or no prior knowledge, a significant proportion of its paradigm may still be correctly predicted. We argue that the hierarchical structures that allow for such default assumptions are associated with a particular kind of morphological complexity, central system complexity, something we discuss in the next section.

7.1 A Three-Dimensional Typology of Complexity

The abstract typology in Chapter 6 addressed the issue of paradigm structure by taking a certain number of inflection classes and determining the different ways in which they can be constituted, considering the difference this makes to morphological complexity. In this section we use them to introduce the three dimensions of complexity that form the basis of our typology:[2]

create a learning model that incorporates phonological similarity and a default rule, the default generalization learner (DGL). The DGL assigns maximal confidence to the default rule that is not sensitive to phonological context. They show that this is a better fit with human responses, in their study of Portuguese conjugation classes, than a model that is based on phonological similarity alone.

[2] Our thinking on these three dimensions of morphological complexity is influenced by work outside of linguistics. The dimensions we consider here draw on Gershenson & Fernández' (2012)

> *Organization:* this is the strength of the rule system and the implicative relations that obtain between members of the paradigm.

In a system that is high in organization, the morphological grammar plays an important role. Paradigmatic structure is easily inferred, because the relationship between the forms and the features gives a clear categorization.

> *Emergent Complexity*: this is the opposite of organization. It is a measure of how much unpredictability there is in the system, and is associated with system entropy.

In a system that is high in emergent complexity the traditional notion of the lexicon and lexical idiosyncrasy plays an important role. Much of the knowledge has to be associated with lexical entries, because it is not predictable.

> *Central System Complexity*: this is the compromise between organization and emergent complexity. Central system complexity is at its highest when the two other types of complexity are in balance.

A system that is high in central system complexity represents an equilibrium between lexical stipulation and grammar. As it involves the intersection of the two other types, this represents the canonical extreme of morphological complexity.

Following the presentation in section 6.1, we explore next the differences between these types through a series of artificial inflection class systems, each made up of four classes. We look at the different ways these systems can be constituted, and what difference this makes to the dimensions of complexity we have identified.

7.1.1 Organization and Emergent Complexity in Grid Systems

Grid systems are those that differ between inflectional classes in every cell, and show absolute inter-predictability between paradigm cells, as in Table 7.1. We refer to this inter-predictability as *organization*. As we noted in section 6.1, a grid system can thus be reduced to one distillation, in the sense of Stump & Finkel (2013) – in effect, a single value – because the morphological expression of every feature value shares the same structure.

discussion of three key notions that appear in the wider literature: emergence, self-organization and complexity. These correspond to our notions of emergent complexity, organization and central system complexity respectively. A key aspect of morphological complexity employed in the current literature is the notion of entropy, which relates most directly to emergent complexity here. Ackerman & Malouf (2013) treat entropy as a measure of I-complexity (integrative complexity), arguing that this type of complexity is kept within manageable bounds.

Table 7.1. *Grid system*

	A	B	C	D
1	a	b	c	d
2	e	f	g	h

Table 7.2. *Cross-classifying (minimal)*

	A	B	C	D
1	a	a	b	b
2	c	d	c	d

Grid systems represent *canonical inflection classes* (Corbett 2009a), or Stump & Finkel's (2013: 81) *maximally transparent inflection classes*, because class distinctions are consistently maintained throughout the paradigm. Understood in terms of allomorphy, a grid system is one in which the number of allomorphs for a given feature value is equal to the number of classes – four in this case. Organization in a grid system is very high, because knowledge of one form gives absolute certainty about the other cells in the paradigm. Conversely, such a system is low in emergent complexity, which is related to uncertainty in the system.

7.1.2 Organization and Emergent Complexity in Cross-Classifying Systems

In a cross-classifying system there is no predictability between the forms found for one value and the forms found for another. If we think of this in terms of allomorphy, the various combinations yield more classes than there are distinct forms. Recall from section 6.1 that we described two variants, which we will describe as *minimal* versus *maximal*. In the minimal system in Table 7.2, our four classes are generated through the minimal means: two values, with two forms per value. Each form for value 1 (a, b) can co-occur with each form for value 2 (c, d). If we think of this in terms of the way lexical sets are divided, for value 1 there are two groupings, one embracing A and B, and one embracing C and D. For value 2 the groupings are different: A and C versus B and D. But if we add a third morphosyntactic value we can generate a third division, one that groups A and D versus B and C. This is illustrated

Table 7.3. *Cross-classifying (maximal)*

	A	B	C	D
1	a	a	b	b
2	c	d	c	d
3	e	f	f	e

Table 7.4. *The elements of Table 7.3 fleshed out exhaustively*

	A	B	D	C	E	F	G	H
1	a	a	b	b	b	a	a	b
2	c	d	c	d	d	c	d	c
3	e	f	f	e	f	f	e	e

by the system in Table 7.3, which we call maximal, because it exploits all the logically possible avenues of cross-classification within the limits (four-classes) we have set out to describe.

In cross-classifying systems, organization is low. In the minimal system in Table 7.2 there is no inter-predictability between the paradigm cells. In the maximal system in Table 7.3, organization is also low, but not as low, because there is some inter-predictability. Not between individual cells, but between sets of cells. For example, although form *a* for value 1 can co-occur with either of the allmorphs for both value 2 (c, d) or value 3 (e, f), if we take the *combined* forms for any two values, then all uncertainty about the form used with the third value is eliminated; thus if value 1 has the form *a* and value 2 has the form *c*, then value 3 must have the form *e*. This difference in structure is perhaps best characterized in terms of another property, which we will call *exhaustiveness*. Exhaustiveness means that every possible paradigmatic combination of forms is realized; in other words, that the cross-classificatory possibilities of the system have been exhausted. For example, if every possible allomorph combination in Table 7.3 were spelled out, the result would be the eight-class system in Table 7.4.

An exhaustive system is also a minimal system, in that the classes are defined by the minimal morphological means, i.e. the smallest number of allomorphs, connected by the densest network of cross-classification. As a result there is no predictability or intermediate structure in the system: the only way to explicitly identify a given class is to list all the forms within it.

Table 7.5 *Hierarchical system*

	A	B	C	D
1	a	b	c	D
2	e	e	f	F

Table 7.6 *Hierarchical system (maximal)*

	A	B	C	D
1	a	b	c	d
2	e	e	f	f
3	g	g	g	g

7.1.3 Organization and Emergent Complexity in Hierarchical Systems

In a hierarchical system, the class associated with one value is a superclass of the classes defined by another value. For example, in Table 7.5, classes A and B, for instance, have different forms for value 1, but they share a form for value 2. They therefore belong to a superclass that uses the same form for value 2, while being sub-classes of that superclass with different forms of their own for value 1. Similarly, class C and D belong to a superclass for value 2, while having their own forms for value 1.

Organization in hierarchical systems is intermediate between grid systems with multiple feature values and minimal classifying systems. The level of organization is similar to maximal cross-classifying systems. If value 1 is given, then value 2 is predictable. On the other hand, value 2 is not predictable from value 1 alone. As with the variations on cross-classifying systems just illustrated, we can distinguish between a minimal and a maximal system by considering a version with three values, in which there is a single form for all four classes (Table 7.6). In this case there are two levels of subsumption: the single superclass of value 3 subsumes the two superclasses of value 2 which each susbsumes two classes in value 1. This system has proportionally more organization than the hierarchical system with only two values, because the form for value 1 predicts the forms for both value 2 and value 3, and the forms for value 2 predict the form for value 3. For the same reasons it is also more organized than the maximal cross-classifying system.

Table 7.7. *Organization in the abstract types*

Type	Organization
Grid	High
Hierarchical (maximal)	High ⇔ Medium
Hierarchical (minimal)	Medium
Cross-classifying (maximal)	Medium ⇔ Low
Cross-classifying (minimal)	Low

7.1.4 *Organization and Emergent Complexity: Summary*

We can summarize organization in the different systems as in Table 7.7. The status of these systems as regards emergent complexity will be the opposite of their behaviour with regard to organization. Cross-classifying systems will tend to be high in emergent complexity, while grid systems will be low in emergent complexity. Hierarchical systems will tend to be intermediate for both. As we shall see in the next section, this balance between organization and emergent complexity is connected with hierarchical systems being high in the third dimension we look at, central system complexity

7.1.5 *Central System Complexity*

Central system complexity rises as the levels of organization and emergent complexity approach each other. It is therefore highest in a hierarchical system, where grammatical inference and lexical stipulation are in balance. This is due to the asymmetry of the implicational relationships between superclasses and subordinate classes: super-classes can be predicted by subordinate classes, but not the other way around. For example, in Table 7.6, the forms for value 1 unambiguously predict the forms for values 2 and 3, but value 3 is completely unhelpful, while the forms for value 2 merely narrow down the form possibilities for value 1. Predictability creates organization, whereas unpredictability brings about emergent complexity; the two together create a high degree of central system complexity. We can see this as representing a competition between the grammar (because predictability implies rules) and the lexicon (because unpredictability implies listed idiosyncrasy).

Grid systems and cross-classifying systems are low in central system complexity, but for different reasons. A grid system is as close as we can get to pure grammar, in that it can be described in terms of rules or mutual implications without any dedicated lexical listing (since any form from the paradigm will do). A cross-classifying system, because of its unpredictability, can make little use of rules or implications, and instead must rely on lexical listing. If it is an exhaustive system, then it consists in effect *SOLELY* of listing, with no need for a grammar as such.

7.2 Measuring the Three Types of Complexity

So far we have shown how the different paradigm types we saw in chapter 6 relate to three different dimensions of complexity. In this section we apply different measures in order to arrive at a more objective way of characterizing the contribution of each dimension to the overall picture of morphological complexity in an inflectional system.

7.2.1 Measuring Organization

Interpredictability between cells is a key element of organization, and it is the hallmark of a grid system. In order to quantify how well one cell can be predicted from the other cells in the paradigm, we turn to Stump & Finkel's (2013: 99) *average cell predictability*. The greater the overall cell predictability, the greater the organization in the morphological system. *Average cell predictability* is measured as in (1).

(1) $$\text{C}\textsc{ell}\text{P}_{<w,\sigma>} = \frac{\left| \left[{}_{m}\left[D_{<w,\sigma>} \right] \right]_{-} < w,\sigma > \right|}{\left| \left[{}_{m}\left[\wp(D_{\mathrm{L}}) \setminus \varnothing \right] \right]_{-} < w,\sigma > \right|}$$

This expresses cell predictability as the proportion of non-empty subsets of cells that uniquely determine (i.e., predict) cell <*w*, σ> compared with all non-empty subsets of cells in the paradigm (the denominator). The equation is refined so that <*w*, σ> itself is excluded from the calculation, because a cell trivially predicts itself. Average cell predictability is the average across all cells in the paradigm.

The figure for average cell predictability rises or falls with the degree of organization. For the four-class systems surveyed in §7.1, the figure is highest for the grid system, namely 1.0, because every cell predicts every other cell.[3] For the minimal cross-classifying system – crucially, an exhaustive system – the average cell predictability is zero. This is because no cell can be predicted by any other cell in the paradigm. (And indeed, if we consider the

[3] Our measurement of cell predictability, using Stump & Finkel's (2013: 368–379) Principal Parts Analyzer, requires us to force values in the grid system to be separate distillations. Stump & Finkel (2013) exclude isomorphic values from their calculations by treating them as the same distillation. Of course, this property of grid systems is precisely why they are so organized.

Table 7.8. *Interim summary: organization*

Complexity Type	Measure	Grid	Hierarchical	Cross-classifying
Organization	Average Cell Predictability	High	High ⇔ Medium	Medium ⇔ Low

larger exhaustive cross-classifying system in Table 7.4, combinations of cells are also unpredictive.) The other types lie in between these extremes, going, in order from highest to lowest: (i) the maximal hierarchical system at 0.56; (ii) the minimal hierchical system at 0.5; and (iii) the maximal cross-classifying system at 0.33. The interim summary for average cell predictability is shown in Table 7.8.

7.2.2 *Measuring Emergent Complexity*

Emergent complexity is associated with entropy. The greater the emergent complexity, the greater the uncertainty associated with the realization of morphology. We use Stump & Finkel's (2013: 337) *IC system's average n-MPS entropy* as a measure for this. In order to be more succinct we will refer to this measure as *n-way entropy*. This measure allows for the natural assumption that some information is already given and it is an average across the whole system. In Stump & Finkel's (2013) terminology *MPS*, or 'morphosyntactic property set' is the combination of morphosyntactic feature values that defines a paradigm cell. The default setting for n-way entropy is 4. What this means is that, for a given lexeme in a particular language, we take every possible combination of 4 cells or less and work out how well each combination of cells predicts every other cell in the paradigm. The results for each combination are then averaged for each cell and across all lexemes. The *n-way entropy* therefore gives us an average measure of uncertainty associated with the typical paradigm cell.

The higher the degree of emergent complexity, the higher the measure for n-way entropy. As expected, this measure tends in the opposite direction from that for organization. Thus, using our exemplary four-class systems from Section 7.1 and assuming two values: (i) the measure for the grid system is 25 (0.25 x 100), (ii) the measure for the hierarchical system is 31 (0.31 x 100); and (iii) for the minimal (and exhaustive) cross-classifying system it is 50 (0.5 x 100).[4] A similar pattern of increasing n-way complexity can be observed if the

[4] We follow Stump & Finkel's (2013: 296) practice of multiplying entropy measures by 100. They argue that it makes the values easier to read. We adhere to this practice in order to emphasize the different nature of the entropy measure from cell predictability, which is the proportions of cells expressed as a number between 0 and 1.

number of values is changed to three: (1) the measure for the grid system is 13 (0.13 x 100); (2) the measure for the maximal hierarchical system is 21 (0.21); and (3) the measure for the maximal (and non-exhaustive) cross-classifying system is 38 (0.38 x 100). For the larger exhaustive cross-classifying system in Table 7.4 the n-way entropy measure is also 50 (the same as for the minimal exhaustive system). These effects relate to the properties we have noted for the different types. In the grid system, implicational relations work in both directions. They are highly organized and therefore low in the uncertainty associated with emergent complexity. For hierarchical systems implications work in one direction and therefore they are not as organized as grid systems. However, they are more organized than cross-classifying systems, where the implications between paradigm cells are either less reliable or non-existent (in the exhaustive systems). With the exception of the exhaustive cross-classifying systems, where there is no intermediate structure that can serve as predictive, the number of values in the paradigm is relevant, because this increases the set of implicative relations. But the basic pattern is clear: emergent complexity is low in grid systems and high in cross-classifying systems, with hierarchical systems somewhere in between. The interim summary for both organization and emergent complexity is given in Table 7.9.

Since organization and emergent complexity are complementary, it would have been possible to use one measure for the two dimensions. We have, however, used two independent measures in order to show that they tend to yield similar results. Where emergent complexity is low, as measured using Stump & Finkel's (2013) n-way entropy, organization tends to be high, as measured using Stump & Finkel's average cell predictability. The grid system is highly structured and can be captured adequately by a grammar that relies heavily on implicative statements with minimal lexical stipulation. A cross-classifying system is, in contrast, heavily reliant on specifying information in the lexical entries. The hierarchical system is intermediate between these two,

Table 7.9. *Interim summary: organization and emergent complexity*

Complexity Type	Measure	Grid	Hierarchical	Cross-classifying
Organization	Average cell predictability	High	High ⇔ Medium	Medium ⇔ Low
Emergent complexity	N-way entropy	Low	Low ⇔ Medium	Medium ⇔ High

representing a balance between the grammar and lexical idiosyncrasy. This property is relevant for our final dimension, central system complexity.

7.2.3 Measuring Central System Complexity

Central system complexity is a balance between the morphological grammar and lexical stipulation. This will also mean that a measure for central system complexity should be highest for the abstract type that is intermediate for organization and emergent complexity, namely the hierarchical type. In contrast, because they represent the different extremes in terms of the trade-off between the morphological grammar and lexical stipulation, the grid and cross-classifying types should be low in terms of the measure for central system complexity.

We use the ratio of optimal dynamic principal part sets to possible sets (*DPP ratio*, for short) as a measure of central system complexity. Stump & Finkel's notion of dynamic principal part usefully captures the property of real morphological systems that they may differ with regard to the number of forms and associated values required to identify inflectional classes. Table 7.10, based on Stump & Finkel (2013: 33), illustrates how dynamic principal parts work. Knowing either of the forms associated with value 1 is sufficent to identify class A or B. Similarly, knowing either of the forms associated with value 2 is sufficient to identify either class C or class D. Knowing the form associated with value 3 is sufficient to identify class F. For class E, one value is insufficient to do this. However, knowing the form associated with value 3 plus one other form (associated with either value 1, 2, or 4) is sufficient. An optimal dynamic principal part analysis, in the terminology of Stump & Finkel (2013: 29–37), has to be both 'adequate' and 'minimal'. To be adequate it has to uniquely determine the other cells in the paradigm. To be minimal there should be no other adequate analysis that relies on a smaller number of values and associated forms. For classes A, B, C, D, and F there is one optimal analysis,

Table 7.10. *Dynamic principal parts*

	A	B	C	D	E*	F
1	a	b	c	c	d	d
2	e	e	f	g	h	h
3	i	i	j	j	j	k
4	l	l	m	m	n	n

Source: Based on Stump & Finkel (2013: 33).

involving one value only. As noted, for class E there are three optimal analyses: {3, 1}, {3,2}, or {3,4}. Stump & Finkel's DPP ratio measure takes into account the size of the minimal analysis for each inflectional class. The ratio of actual to possible analyses of that size is calculated for the class. The DPP ratio measure is the average across all classes. We can illustrate with Table 7.10. Class A has a dynamic principal part analysis of size one (one pairing of a value and a form). There are four possible analyses of size one, but only one is optimal (value 1). The ratio for class A is therefore 25 percent. The ratios for classes B, C, D, and F are also 25 percent. Class E, on the other hand, has three optimal analyses of size two. There are six ways of selecting two from four. So the ratio of actual to possible for class E is 50 percent (3/6). The overall DPP ratio for the system in Table 7.10 is the average of the ratios for the six classes, 29.17 percent.

Central system complexity arises from the compromise between lexical stipulation and the morphological grammar. Stump & Finkel (2013: 330) note that a small DPP ratio is an indicator of high complexity. We interpret a low DPP ratio to be an indicator of high central system complexity. This is because systems that can be accounted for with a mixture of lexical stipulation and morphological rules will underutilize the space of possible analyses. For instance, the structure of hierarchical systems (section 7.1.3), being uni-directional, will not exploit the full space of possible principal parts.

Low central system complexity, indicated by a high DPP ratio, can arise for two different reasons. At one extreme, a system has many possible optimal principal parts, because of the many implicative relations, and all of these can be used (a system based on a pure morphological grammar). This is true of the grid system in Table 7.1. For the grid system, an optimal dynamic principal part analysis is always of size one; one cell is required to predict the other cells in the paradigm. There are many optimal principal part analyses, and they all work. The DPP ratio for a grid system is 100 percent (low central system complexity). Conceptualized as the balance between morphological grammar and lexical stipulation, all the weight is placed on the morphological grammar, with little lexical stipulation required.

At the other extreme a system based purely on lexical stipulation has only one possible optimal dynamic principal part analysis, and this is used. The exhaustive cross-classifying system in Table 7.4, for instance, can only work through knowledge of the forms for all three values (the complete paradigm). There are no super-classes for which generalizations can be made by the morphological grammar. Instead, knowledge of most of the paradigm, or all of it, is required by the cross-classifying system. This means that the actual is the same size as the possible. For an exhaustive cross-classifying system, the DPP ratio is 100 percent because the entire paradigm must be lexically stipulated, so there is only one set of optimal principal parts. For the non-exhaustive maximal cross-classifying system in Table 7.4 the DPP ratio is also 100 percent. Here the optimal principal part set consists of any two cells (out of a total of three), and every combination of two cells is equally good, yielding full exploitation of the optimal principal part set and a DPP ratio of 100 percent. Conceived in terms of the opposing extremes of lexical stipulation and morphological grammar, a high DPP ratio is a measure of the consistency of the system; a high measure indicates that structures of a similar size behave in the same way. If, as with a consistent morphological grammar, one form can be inferred from another, then this is consistently true for every pairing of forms. If we need to lexically stipulate more than one form to know the full paradigm, then this is consistently true throughout the system. In contrast, a system with high central system complexity arises from the mixture of the morphological grammar and lexical stipulation.[5]

7.2.4 Summary for Grid, Cross-Classifying, and Hierarchical Systems

The different abstract types were presented to Stump & Finkel's Principal Parts Analyzer.[6] Each of the three measures in Table 7.11 characterizes different

Table 7.11. *Measures for the abstract types*

	Organization	Emergent Complexity	Central System Complexity
	Cell Predictability	*n-way entropy*	*DPP Ratio*
C-C (min)	0	50	100%
C-C (exh.)	0	50	100%
C-C (max)	0.33	38	100%
Hierarchical	0.5	31	50%
Grid[a]	1.0	25	100%

[a] The score for grid here is the result of forcing separate distillations in the PPA Analyzer.

[5] As with the n-way entropy measure, the DPP ratio is calculated by specifying a maximum size for the dynamic principal parts analysis, set to 4 by default. For an exhaustive classifying system, the size needs to be set higher than 4, if the number of values is greater than this, because only knowledge of the complete paradigm is sufficient.

[6] The Principal Parts Analyzer is available from www.cs.uky.edu/~raphael/linguistics/analyze.html and is described in Stump & Finkel (2013: 368–379).

Table 7.12. *Summary: organization, emergent complexity, and central-system complexity*

Complexity Type	Measure	Grid	Hierarchical	Cross-classifying
Organization	Average cell predictability	High	High ⇔ Medium	Medium ⇔ Low
Emergent complexity	N-way entropy	Low	Low ⇔ Medium	Medium ⇔ High
Central System Complexity	DPP Ratio	Low	Medium ⇔ High ⇔ Medium	Low

properties of a morphological system. A high cell-predictability measure, used for organization, indicates that the morphological grammar prevails. It also indicates that lexical stipulation is low. As a separate check on this, we use the n-way entropy measure for emergent complexity. Where this is high, it shows that more information has to be stipulated lexically, because it is not readily predictable. So the systems with higher measures for emergent complexity tend to score low for organization. As discussed in the previous section, high central-system complexity is indicated by a low DPP ratio. We can see that low central-system complexity is associated with systems that are either low in organization and high in emergent complexity (cross-classifying systems) or high in organzation and low in emergent complexity (grid systems).

Taking these points into account, we arrive at the basic summary of the abstract types in Table 7.12. These are abstract types, of course, and real morphological systems will almost certainly involve a combination of these. The combination of one type with another can fundamentally alter the nature of the relationship, as we saw in our discussion in Chapter 6 of how we arrive at these systems. As we will see in our case study in section 7.3, central system complexity can be higher than we observe for the abstract systems presented, because the measures are dependent on the number of classes. However, for systems with a similar number of classes, a similar pattern emerges to that presented in Table 7.12. In such a case, what we observe is a lot of structure that is intermediate between lexical stipulation on the one hand and a morphological rule system that generalizes without exception. We now consider how a real system – one of considerable prima facie complexity – relates to these measures.

7.3 Case Study: Tlatepuzco Chinantec

Tlatepuzco Chinantec is a Chinantecan (Oto-Manguean) language spoken in the village of San Pedro Tlatepuzco in Oaxaca state, Mexico, by approximately 25,000 people (Feist and Palancar 2015). Data from this language were presented in section 6.4 in connection with the interaction of parallel inflection class systems. Here we concentrate solely on the tonal realization of subject person-number and TAM (most verbs in fact do not have a stem alternation). This system is striking both for the proliferation of inflection classes (over 80 different types) and the lack of any obvious structure to the patterning, making it an ideal subject for closer study and analysis.

A salient feature of this system is the restricted number of morphological exponents (five tones: 1 low, 2 mid, 12 low-mid, 3 high and 13 low high) and their seeming independence from any even approximately consistent function. To get a flavour of the way this system is constituted, consider the two verbs in Table 7.13.[7] For both of them, the paradigm is composed of all five available tones, but apart from the third-person future and second-person completive (with tone 1), the distribution of these tones in the paradigm is completely different. Multiply this sort of thing out several times and we have the dozens of inflection classes that characterize Tlatepuzco Chinantec. They make up a mixed system of allomorphic and distributional inflection classes that defy easy categorization in the terms we have discussed so far. We show, however, that a somewhat more abstract view of how the paradigm is composed reveals elements of structure that make Tlatepuzco Chinantec less of a typological outlier than first impressions might suggest.

Table 7.13. *Representative Tlatepuzco Chinantec verb paradigms*

a. 'take out'					b. 'follow'			
	1SG	1PL	2	3	1SG	1PL	2	3
PRS	dsi^{12}	dsi^{2}	dsi^{12}	dsi^{2}	hen^{2}	hen^{12}	hen^{2}	hen^{12}
FUT	dsi^{13}	dsi^{3}	dsi^{13}	dsi^{1}	hen^{3}	hen^{13}	hen^{3}	hen^{1}
CPL	dsi^{1}	dsi^{3}	dsi^{1}	dsi^{2}	hen^{2}	hen^{13}	hen^{1}	hen^{1}

[7] This represents a somewhat abbreviated picture of the total paradigm, but one which is sufficient to portray the inflection classes. There are further TAM distinctions made by transparent prefixation to the forms shown here, and they are consistent across the lexicon. There are also tonally distinguished directional categories, which, however, do not appear to be available to every verb, and are, in any case, not included in Merrifield & Anderson's (2007) tabulation of inflectional types.

There are 777 Tlatepuzco verbs in the dataset available online (Feist and Palancar 2015). Of these, 618 verbs have full person-number paradigms (the rest are verbs whose subject is always inanimate and hence lack first and second person forms). We restrict our investigation here to these 618 verbs. We will see in section 7.3.1 that the Tlatepuzco verbal system exhibits the interesting property of having very high central system complexity, while being intermediate between the extremes of a grid system and a cross-classifying system in terms of organization and emergent complexity.

7.3.1 Organization, Emergent Complexity, and Central-System Complexity

We saw that Tlatepuzco Chinantec makes use of a system of five tones. The Tlatepuzco Chinantec data suggest at first blush that any tone can occur with any other tone so that it has the properties of a cross-classifying system. However, on closer inspection, its level of organization is intermediate between a cross-classifying system and a grid system. In particular, if we assume an intermediate level of structure involving cohorts of cells (which we call inflectional series, following Palancar 2014), the system has a strong admixture of hierarchical structure, increasing its central system complexity.

In order to assess the Tlatepuzco Chinantec system according to the measures outlined in section 7.2, we will need to compare like with like as much as possible: recall that the figures in given in Table 7.11 were for systems with only four inflection classes, whereas the Tlatepuzco Chinantec dataset contains 84 unique classes. Therefore, we created an abstract exhaustive cross-classifying system containing 6 values and 2 allomorphs, resulting in 64 classes (2^6). We also created a hierarchical system that resulted in 64 classes, and a grid system that contained 64 classes, all with unique allomorphs.[8]

Table 7.14 shows that the Tlatepuzco Chinantec dataset is closest to the larger hierarchical system, but is less organized. The Tlatepuzco system also has greater emergent complexity. As expected, both the cross-classifying and grid systems have a 100 percent DPP ratio and, therefore, low central system complexity. Tlatepuzco Chinantec and the larger hierarchical system, in contrast, have low DPP ratios and, therefore, lie in the medium to high range for central-system complexity (in accord with Table 7.12). Tlatepuzco Chinantec and the larger hierarchical system also lie in the medium to high range for

[8] For the calculation of the DPP ratio in the cross-classifying system the size of the DPP analysis is set to 6 (i.e., the whole paradigm), because no principal part set can be calculated from less than the complete set of cells.

Table 7.14. *Tlatepuzco's position*

	Organization	Emergent Complexity	Central System Complexity
	Cell Predictability	*n-way entropy*	*DPP Ratio*
Cross-classifying (64 classes)	0	50	100%
Tlatepuzco Chinantec (84 classes)	0.481	17	7.29%
Hierarchical (64 classes)	0.689	3	16.67%
Grid (64 classes)	1.0	0	100%

organization, and correspondingly in the low to medium range for emergent complexity (again, in accord with Table 7.12). These figures show that the Tlatepuzco verbal system behaves very differently from a cross-classifying system, and it is closer in a number of respects to a hierarchical system. As we shall see, the Tlatepuzco lexicon contains many mid-scale structures that can be characterized as hierarchical, and its higher central system complexity is, therefore, to be expected.

7.3.2 Intermediate Structure in Tlatepuzco Chinantec

The measures given in Table 7.14 suggest that there is some structure to the Tlatepuzco Chinantec system, but they do not tell us how to interpret that observation. The clearest indication of a *lack* of structure is for a system to be exhaustively cross-classifying. But since that would require Tlatepuzco Chinantec to have many times more inflection classes (244,140,625) than there are verbs in the language, it could be that this appearance of structure is just an illusory by-product of the size of the verbal lexicon. Simply scanning the wealth of tonal patterns brings no obvious resolution of the question. But a clearer picture emerges if we step back from the individual paradgim cells and start looking instead at clusters of cells. The impetus to do this is the very ambiguity of the inflectional exponents. The Tlatepuzco Chinantec tonal paradigm is a prime example of a mixed allophonic and distributional class system, with class differences defined both by different tonemes, and different distributions of the same toneme. The latter, in particular, means that no tonal exponent of inflection has any identifiable meaning anywhere in the system. That in itself should not perplex us too much, since the independence of form and meaning is the topic of this very book. But it is rare that we find no hint, however vague or faint, of identifiable form-meaning pairings, if only

Table 7.15. *Competing analytic choices*

a.

	1SG	1PL	2	3
PRS	a	b	c	d
FUT	e	f	g	h
CPL	i	j	k	l

b.

	1SG	1PL	2	3
PRS	a	b	c	d
FUT	e	f	g	h
CPL	i	j	k	l

as a diachronic memory. The clue here is to separate out the features that are cumulated in the paradigm cells – namely, person-number on the one hand and TAM on the other. This yields two analytic choices, schematically illustrated in Table 7.15. One is to treat the paradigm as consisting of three TAM forms, with PRS having the value {a,b,c,d}, FUT {e,f,g,h}, and so on. The other is to treat the paradigm as consisting of four person-number forms, as in Table 7.15b, where 1SG has the value {a,e,i}, 1PL has the value {b,f,j}, and so on. On this construal, the 'form' is actually a set of forms. Following Palancar 2014 (based in turn on the analysis in Merrifield & Anderson 2007) we call this compound form an *inflectional series*.

The two alternative analyses yield very different results. TAM-based inflectional series yield a number of distinct allomorphs without lending any additional clarity to the system. But with person-number-based inflectional series, patterns start to emerge. For one thing, there are at least some dedicated forms; for example, the series {2,2,2}, {2,1,2}, {12,1,1} are almost invariably an exponent of third person, {12,13,12}, while {2,3,12} and {2,3,1} are exponents of second person, and so forth (Baerman & Palancar 2014: 56).[9] The second observation is that the predictiveness and predictability of person-number-based series is much greater than that of TAM-based series. This suggests that if we want to understand the structure of the Tlatepuzco Chinantec paradigm, we should treat it as one which marks person-number, having absorbed the TAM distinctions.

On this analysis, the Tlatepuzco Chinantec inflectional system is deeply coloured by hierarchical structure. This is largely a result of the way distributional classes are configured. To illustrate this, let us take inflectional series 5 (referred to here using the system outlined by Palancar 2014) as an example. This consists of tone 12 in the present, tone 13 in the future and tone 1 in the completive. For some verbs it is used in the 1SG (Table 7.16), for others both 1SG and second person (Table 7.17), others add 1PL to this (Table 7.18), and

[9] We provide a list of Palancar's (2014) inflectional series in an appendix.

Table 7.16. *Tlatepuzco Chinantec 'covet'*

	1SG	1PL	2	3
PRS	tión^{12}	tión^{12}	tión^{12}	tión^{12}
FUT	tión^{13}	tión^{13}	tión^{13}	tión^{1}
CPL	tión^{1}	tión^{13}	tión^{13}	tión^{1}

Table 7.17. *Tlatepuzco Chinantec 'receive'*

	1SG	1PL	2	3
PRS	ʔien^{12}	ʔien^{12}	ʔien^{12}	ʔien^{12}
FUT	ʔien^{13}	ʔien^{13}	ʔien^{13}	ʔien^{1}
CPL	ʔien^{1}	ʔien^{13}	ʔien^{1}	ʔien^{1}

Table 7.18. *Tlatepuzco Chinantec 'defecate'*

	1SG	1PL	2	3
PRS	tø^{2}gugh^{12}	tø^{2}gugh^{12}	tø^{2}gugh^{12}	tø^{2}gugh^{12}
FUT	tø^{2}gugh^{13}	tø^{2}gugh^{13}	tø^{2}gugh^{13}	tø^{2}gugh^{12}
CPL	tø^{2}gugh^{1}	tø^{2}gugh^{1}	tø^{2}gugh^{1}	tø^{2}gugh^{12}

Table 7.19. *Tlatepuzco Chinantec 'disappear'*

	1SG	1PL	2	3
PRS	kionʔ^{12}	kionʔ^{12}	kionʔ^{12}	kionʔ^{12}
FUT	kionʔ^{13}	kionʔ^{13}	kionʔ^{13}	kionʔ^{13}
CPL	kionʔ^{1}	kionʔ^{1}	kionʔ^{1}	kionʔ^{1}

still others use this series for all persons (Table 7.19). Its distribution across different lexemes is characterized by spreading syncretism. If we treat the different distributions of series 5 as different inflection classes, the result is the near-perfect hierarchical system in Table 7.20, representing 246 lexemes (out of the 618 in the dataset).

Other such hierarchical structures occur with other series as well, sometimes following a different paradigmatic patterning. For example, series D (tone 12

Table 7.20. *Hierarchical structure for inflectional series 5*

	type I (9 lexemes)	type II (2 lexemes)	type III (100 lexemes)	type IV (132 lexemes)	type V (3 lexemes)
1SG	series 5	series 5	series 5	series 5	
2	series 5	series 5	series 5		series 5
1PL	series 5	series 5			
3	series 5				

Table 7.21. *Tlatepuzco Chinantec 'fall out'*

	1SG	1PL	2	3
PRS	kanʔ^{12}	kanʔ^{12}	kanʔ^{12}	kanʔ^{12}
FUT	kanʔ^{13}	kanʔ^{13}	kanʔ^{13}	kanʔ^{1}
CPL	kanʔ^{1}	kanʔ^{13}	kanʔ^{1}	kanʔ^{1}

Table 7.22. *Tlatepuzco Chinantec 'covet'*

	1SG	1PL	2	3
PRS	tión^{12}	tión^{12}	tión^{12}	tión^{12}
FUT	tión^{13}	tión^{13}	tión^{13}	tión^{1}
CPL	tión^{1}	tión^{13}	tión^{13}	tión^{1}

Table 7.23. *Tlatepuzco Chinantec 'stammer'*

	1SG	1PL	2	3
PRS	cø^{2}cáng^{12}	cø^{2}cáng^{12}	cø^{2}cáng^{12}	cø^{2}cáng^{12}
FUT	cø^{2}cáng^{13}	cø^{2}cáng^{13}	cø^{2}cáng^{13}	cø^{2}cáng^{12}
CPL	cø^{2}cáng^{13}	cø^{2}cáng^{13}	cø^{2}cáng^{13}	cø^{2}cáng^{12}

in the present and 13 in the future and completive) involves the spreading of a 1PL form, rather than a 1SG form as with series 5. Thus, some verbs have series D for 1PL only (Table 7.21), some have it for both 1PL and second person (Table 7.22), and one verb has it for 1PL, second person, and 1SG (Table 7.23). (Note that some verbs, such as 'covet' participate in more than one hierarchical structure at the same time.)

The resulting hierarchical inflection class structure is shown in Table 7.24.

Table 7.24. *Hierarchical structure for inflectional series D*

	type a (1 Lexeme)	type b (15 Lexemes)	type c (140 Lexemes)
1PL	series D	series D	series D
2	series D	series D	
1SG	series D		

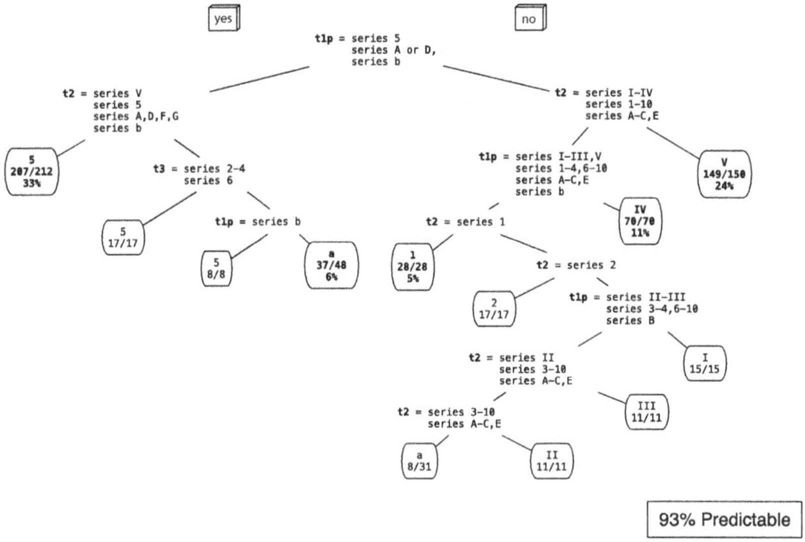

Figure 7.1. *Predicting the inflectional series for the first singular.*

The structuring we have presented for series 5 and series D is repeated across the inflectional series in the Tlatepuzco lexicon. Given sufficient information, this makes it possible to infer the correct inflectional series and, therefore, correct realization for the majority of the lexicon. Exponence in the first person singular is the most predictable (93% of the lexicon), then the first person plural (88% of the lexicon), followed by the second person (84%), with the third person being the least predictable (65%). This is shown for the first-person singular and first-person plural by the classification-trees analyses presented in Figures 7.1–7.2.[10] As their name suggests, classification trees are a method for data classification (Baayen 2008: 148). They provide decision rules based on a

[10] Each of the trees has been created and pruned according to the method outlined in Baayen (2008: 148–154).

set of predictors. The algorithm used to create the tree chooses the most helpful predictors, in our case for determining the inflectional series associated with a particular person and number. In Figure 7.1 the classification tree predicts the inflectional series of the first-person singular, having been presented with the inflectional series of the first-person plural, the second person, and the third person as possible predictors.

The best predictors are given at each node in the tree. If there is a match with any of the predictors, we move down the left branch of the tree. If there is not, we move down the right branch. We keep doing this until we reach a terminal node that gives a predicted value and tells us how many of the observations are correctly predicted. As noted, Figure 7.1 predicts the inflectional series for the first-person singular. The first set of predictors are associated with the series for the first-person plural (t1p).[11] If the inflectional series 5, A, D, or b are used for the first plural, we move down the left branch of the tree. If the second person (t2) is realized by series V, series 5, A, D, F, G, or b, then the inflectional series for the first-person singular is predicted to be series 5. This is correct for 207 out of 212 observations that match with the predictors that directed us down the left branch. The percentage figure given is the proportion of the whole lexicon (618) that the correct observations account for (207/618).

Note that the hierarchical relationship between the first-person plural, second person, and first-person singular that we have observed for series 5 in Table 7.20 is contained within the set of predictors for the classification tree. The presence of series 5 in the first-person plural and second person entails series 5 in the first-person singular. This is one of the reasons that first-person plural and second person serve as good predictors of the first-person singular in Figure 7.1.

In Table 7.24 we saw that series D in second person or first-person singular implies series D in the first-person plural. We expect series D to be predictable for the first-person plural. This is confirmed in Figure 7.2. Here the initial set of predictors requires that either series I to IV, 1 to 10, or B, D or series a are used to express the first-person singular (t1s). Because the pattern we are interested in has series D in the first-person singular, we move down the left branch. The next set of predictors requires examination of the second person (t2), but they do not mention series D. So inflectional series D in second singular does not satisfy the predictor and we move down the right

[11] In the tree, the variables associated with each person and number are: t1p (first person plural), t1s (first person singular), t2 (second person), t3. The initial t is there to indicate that these relate to the inflectional series for tone. The levels of the variable are the different inflectional series.

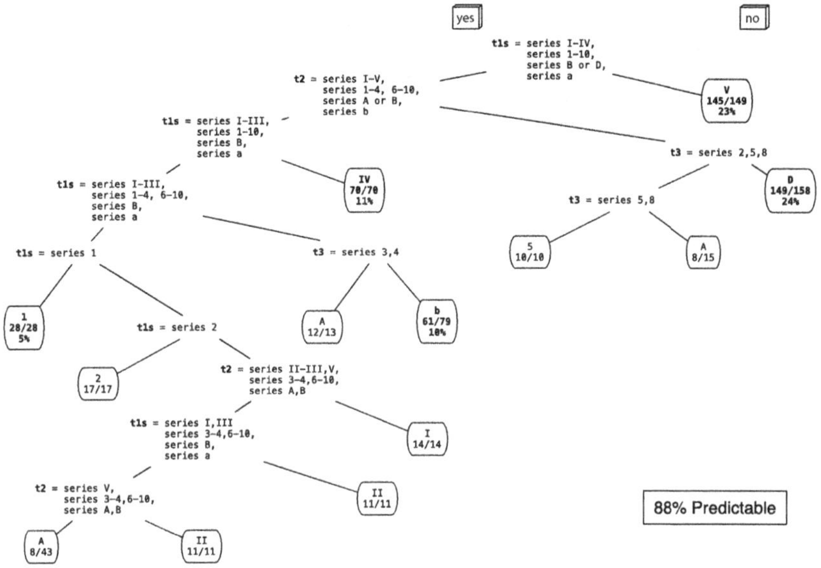

Figure 7.2. *Predicting the inflectional series for the first person plural.*

branch. Again, the predictors do not mention series D. So the right branch is traversed again, and it is stated that, of the observations that match the first predictor and fail to match the next two, 149 out of 158 observations have pattern D in the first person plural. The correct observations account for 24 percent (149/618) of the total lexicon. The hierarchical structure we showed earlier for series D accounted for 156 lexemes, all of which had series D in the first-person plural. The difference of 7 between the 149 correct observations in Figure 7.2 and the 156 lexemes in our hierarchical model in Table 7.24 is accounted for by the elimination of 6 lexemes with series B in the second person, and 1 lexeme with series 2 in the second person. But these still conform to the hierarchical structure.

We have presented an analysis of the Tlatepuzco Chinantec data based on an intermediate structure, the inflectional series. The analysis shows two key things about the Tlatepuzco Chinantec verbal lexicon. First, inflectional series provide a more reliable association with person and number. Second, the inflectional series can often be related in a hierarchical structure, and the classification trees confirm that there are very effective implicational relations between them. In systems like this, there is a compromise between lexical stipulation and the morphological grammar, and this is associated with high central-system complexity, as discussed in section 7.2.

In the next section (section 7.3.3) we concentrate on what needs to be lexically stipulated, showing that a static principal-parts system that exhausts knowledge of values and forms (value maximization) works effectively with the inflectional series system, illustrating the compromise between lexical stipulation and the morphological grammar.[12] In section 7.3.4 we present a complementary analysis that illustrates the power of defaults when used as part of the morphological grammar. In particular, default classes define the tones for sets of cells, and, therefore, implicational relations between them. A default assignment system taps into these classes. Using all four, three, two, one, or none of the value-maximizing principal-part set we show that the model performs well on two measures; (1) cell success (correctly predicting forms) and (2) lexeme success (correctly predicting complete paradigms). The set of analyses we present is complementary, illustrating the role of lexical stipulation, such as principal parts and rule systems. These need to be understood both in terms of implicational relations and defaults. We now turn to consider what needs to be lexically stipulated in Tlatepuzco Chinantec.

7.3.3 Principal Parts and Value Maximization: Lexical Stipulation Facilitates Grammatical Inference

As we have seen, the weak relationship between exponents and feature values means that knowing the form associated with a single cell of the paradigm is insufficient to predict the other forms in the Tlatepuzco Chinantec verbal system. This problem is mitigated substantially if the three-cell inflectional series based on person and number is adopted as an organizing structure. This is an important analytical step for understanding the Tlatepuzco Chinantec system. But is there anything else underlying the system that supports the inflectional series approach? It turns out that the distribution of forms is such that they maximize knowledge of the values. Value maximization, as defined in (2), is a property of the Tlatepuzco Chinantec system.

(2) Value maximization
The most informative set of principal parts is distributed in such a way that it exhausts the set of values that define the paradigm. It should do this using a minimal set of forms.

Three features define the paradigm: person, number, and TAM. Person and TAM have three values, whereas number has two. Where every combination of the features is possible, the number of forms required is the same as the

[12] The idea that the static principal part system for Tlatepuzco Chinantec maximizes knowledge of the values is due to Olivier Bonami.

Table 7.25. *The feature–value combination that is the most predictive*

	1SG	1PL	2	3
PRS				
FUT				
CPL				

cardinality of the largest feature. Here, that would be three. In the Tlatepuzco Chinantec system, however, number is distinguished only for first person. This means that the minimal set of forms required is four, because number creates an additional distinction within the first person only. In Table 7.25 we see the combination of four paradigm cells that is the most predictive of the eight remaining cells in the paradigm.

There are 495 ways of choosing four cells from twelve. Of these 495 possible combinations, the one in Table 7.25 has the lowest average conditional entropy for predicting the other cells in the paradigm.[13] It obeys (2), because the four cells exhaust the set of values. That is, each value in the three sets of features {SG,PL}, {1,2,3} and {PRS, FUT, CPL} is associated with one of the four cells in Table 7.25. It also turns out that, out of the 495 possible combinations of four cells, the best six combinations with the lowest average conditional entropy for predicting the other cells all obey (2). Conversely, of the 495 possible combinations of four cells, there is no combination in the bottom 200 (i.e., those which have the highest conditional entropy as predictors) that obeys (2).[14] If we consider combinations of less than four cells, then those with the lowest conditional entropy as predictors are ones that maximize the set of values as much as possible. For three-cell combinations, for instance, the best performer is {1 PRS, 1PL FUT, 3 CPL}. Here, no person, number, or TAM value is repeated. Naturally, one has to be missing, given the number of cells in the combination. A similar statement could be made for the best two-cell combination {1SG FUT, 1PL PRES}.

[13] The number is 0.069. In contrast with the n-way entropy measure discussed earlier, the number reported here is not multiplied by 100. It was calculated using a program created by Olivier Bonami and further adapted by Sacha Beniamine.

[14] As a predictor of the other cells in the paradigm (measured using average conditional entropy) the worst performing combination of four cells is {1SG FUT, 2 FUT, 1PL CPL, 1PL FUT}. Note the absence of the third person and the present from this combination of feature values. Of the combinations where feature orthogonality is maximized the worst performing is {1SG FUT, 2 FUT, 1PL CPL, 3 PRS}. It is the 293rd most predictive combination. Note the significant overlap with the worst performing combination overall.

Given knowledge of the four cells in Table 7.25, a speaker of Tlatepuzco should be able to determine the tone exponent for any other cell in the paradigm by making using of three types of information: (a) knowledge of the possible inflectional series associated with the person and number value; (b) knowledge of how the inflectional series of one or more person and number values predicts the inflectional series of another person and number value; (c) relationships between paradigm cells.

Let us return to the verb *ʔien*[12] 'receive' that we saw earlier in Table 7.17 to see how this is achieved. The information in Table 7.26 would be available. The inflectional series associated with tone 12 in the first-person singular present are given in Table 7.27, with their Palancar series label shown in parentheses where it is relevant for our discussion. The inflectional series associated with tone 13 in the first-person plural future are given in Table 7.28. The inflectional series associated with tone 1 in the second-person completive are given in Table 7.29. The inflectional series associated with tone 1 in the third-person future are given in Table 7.30.

A speaker of Tlatepuzco knows that the realization of the full paradigm must be a combination of the possible inflectional series in Tables 7.27–7.30.

Table 7.26. *The predictive tones for ʔien*[12] *'receive'*

	1SG	1PL	2	3
PRS	ʔien[12]			
FUT		ʔien[13]		ʔien[1]
CPL			ʔien[1]	

Table 7.27. *Inflectional series for first-person singular predicted by tone 12 in the first-person singular*

Inflectional Series (1 PRS, 1SG CPL, 1SG FUT)	Number of Lexemes
12,1,13 (Series 5)	243
12,1,1 (Series 1)	28
12,12,12	12
12,12,1	2
12,1,2	2
12,2,1	1
12,13,13 (Series D)	1

Table 7.28. *Inflectional series for first-person plural predicted by tone 13 in the first-person plural*

Inflectional Series (1PL PRES, 1PL CPL, 1PL FUT)	Number of Lexemes
12,13,13 (Series D)	155
13,13,13	74
12,1,13 (Series 5)	11
2,13,13	7
2,1,13	1

Table 7.29. *Inflectional series for second person predicted by tone 1 in the second-person completive*

Inflectional Series (2 PRS, 2 CPL, 2 FUT)	Number of Lexemes
12,1,13 (Series 5)	114
12,1,1 (Series 1)	28
2,1,3 (Series C)	18
1,1,1	11
2,1,1	7
2,1,2	4
12,1,2	2
2,1,13	1

Table 7.30. *Inflectional series for third person predicted by tone 1 in the third-person future*

Inflectional Series (3 PRS, 3 CPL, 3 FUT)	Number of Lexemes
12,1,1 (Series 1)	240
2,2,1 (Series 6)	34
1,1,1 (Series III)	31
12,2,1 (Series 8)	18
2,1,1 (Series 4)	10
3,3,1 (Series 7)	3
12,12,1 (Series 9)	2
12,2,1 (Series 3)	1

How does a speaker of Tlatepuzco infer what the correct inflectional series for the first-person singular is? There is a relationship of type (c), between paradigm cells, that is very helpful.

(3) If the first-person plural future has tone 13 and the first singular present has tone 12, then the first-person singular future will be tone 13. (This has one exception.)

This means that the first-person singular must be either inflectional series 5 or D. Furthermore, we are also helped by type (b) knowledge, relationships between inflectional series, in choosing between these two options: we saw from the hierarchical organization associated with inflectional series 5 and D in Tables 7.20 and 7.24 that it is most likely that the first singular will be series 5, because there is only one lexeme with series D for the first singular.

(4) Inference: the inflectional series associated with *ʔien*^{12} in the first-person singular is series 5.

Further knowledge of type (b) is particularly important for inferring the series associated with second person, and organization plays an important role because competing hierarchical structures allow the form to be inferred straightforwardly. The hierarchical structure we have observed earlier for inflectional series 5 in Table 7.20 indicates that it is most likely that the first singular will be series 5 on its own (100 lexemes), or that the second person will be series 5 in addition to the first singular (132 lexemes). As we can see in Table 7.31, the hierarchical structure associated with series 1 allows us to rule out series 1 for the second person, because this occurs only when the first person singular is series 1 (type α).

Type (b) knowledge (i.e., of the relationship between inflectional series) helps determine the realization of the second person. Of the remaining possibilities for the second person in Table 7.29, none of them occurs with series 5

Table 7.31. *Series 1 hierarchical structure facilitates inference*

	type α 28 Lexemes	type β 1 Lexeme	type γ 211 Lexemes
3	series 1	series 1	series 1
1PL	series 1	series 1	
1SG	series 1		
2	series 1		

in the first person singular, with the exception of series C. There are, however, only two lexemes with series 5 in the first person singular that have series C in the second person, and one of these occurs with a third-person series (series 2) that is not included in Table 7.30 and, therefore, not possible.

(5) Inference: the inflectional series associated with *ʔien*^{12} in the second person is series 5.

Inference of the series for the third person and first person plural is straightforward. For the third person in Table 7.30, the series III, 4, 7, and 9 do not occur with series 5 in the second person and the first person singular. The remaining third person series that do (1, 6, 8, and 3) all occur with series D in the first-person plural, with the exception of one lexeme (series 8 in the third person and series 5 in the first-person plural). It can, therefore, be inferred that the first-person plural is realized by series D.

(6) Inference: the inflectional series associated with *ʔien*^{12} in the first person plural is series D.

For the third person (Table 7.30), series 1 is the most reliable inference. There are only three lexemes that have series 3 in the third person with the combination of series that we have inferred for the other persons, and there is only one that has series 6 with that combination.

(7) Inference: the inflectional series associated with *ʔien*^{12} in the third person is series 1.

We can see that the inflectional series provides a useful structure that allows the speaker of Tlatepuzco to make use of implicational relations and organize their knowledge of morphology. The hierarchical organization that can be observed in the way that many inflectional series are related for the person feature provides a highly predictive structure.

What we have done here essentially is to look at the Tlatepuzco Chinantec data from the perspective of different morphological scales. Value maximization in (2) involved a different level of structure (individual paradigm cells) compared with the intermediate-scale inflectional series. In adopting an approach such as this, using overlapping generalizations, we are taking seriously the view articulated in Blevins, Ackerman, Malouf, & Ramscar (2016) that what they term the 'continuity hypothesis' would lead us to overlook significant aspects of morphological systems. The continuity hypothesis is the reductive assumption that the properties of the component parts are contained within the larger scale object. As we have seen here, however, knowledge of different structures provides useful generalizations that would be lost if we opted for a single morphological scale. Implicational relations can be found within these different levels of structure and between them. But we also need to take into account differences in the strength of implications. In their study of the differing roles of type frequency and implicational structure, Sims & Parker (2016) note that, for languages like Tlatepuzco Chinantec, implicational structure does more work if weighting minimizes the influence of small classes.[15] In the next section, we show exactly how default class assignment exploits the stronger implicational relations of the default classes, and that even where a lexeme is assigned to a higher class in the absence of sufficient information, a significant proportion of the paradigm is correctly predicted, because of the stronger implicational relations within the default classes.

7.3.4 Combining Defaults and Implicative Relations

So far we have considered the most predictive static principal parts in terms of a composite analysis that considers individual cells and the larger scale inflectional series system. A key issue with the Tlatepuzco Chinantec system is that there are many small classes that are hard to characterize. Stump & Finkel (2013: 225) argue that marginal inflectional classes are the ones that detract most strongly from the predictability of other inflectional classes (the Marginal Detraction Hypothesis). A default inheritance system is ideal for treating a system with this property, because it allows for the smaller classes to override rules of exponence while still inheriting key generalizations, just as, for a non-linguistic system, default inheritance allows us to deal with penguins by overriding general statements about flying while still allowing for the inheritance of the property of having beaks and feathers, for instance. We use the set of four principal parts from the previous section, but we allow for differing degrees of knowledge about them, from none up to the complete set of four. We provide results that show that the smaller classes still share significant portions of the paradigm inherited from the default class. This means that even where a lexeme is incorrectly assigned to a higher, default class, because of insufficient information, there will still be a significant portion of its paradigm that is correctly predicted.

The default inheritance model that we present here consists of two key components:

(8) a. A hierarchy of inflectional classes that associates tones with values.

[15] Sims & Parker (2016) refer to the language as Palantla Chinantec, as do Stump & Finkel (2013).

b. A set of default assignment rules that assign a lexeme to an inflectional class in (8a) depending on the information available.

The hierarchy (8a) represents default assumptions about how tones are associated with values. Classes in the inflectional class hierarchy define complete paradigms. Sub-classes inherit from higher classes but override statements that associate tones with values. We will return to this relationship between higher classes and sub-classes when considering the effects of assignment to a default class on the basis of insufficient information.

Turning from the inflectional class hierarchy we focus on the assignment rules in (8b). These are a partial representation of the kind of knowledge required to infer the inflectional class for a Tlatepuzco Chinantec verb, based on the most predictive combination of cells, discussed in section 7.3.3 The assignment rules in (8b) will direct a lexeme to a point on the hierarchy in (8a). If a lexeme is exceptional in its behaviour, more information will be required to direct it to the correct class. We will illustrate how this works using the lexeme *ʔien*^{12}.

The assignment rules are represented in DATR (Evans & Gazdar 1996) and take the following form (9), with ellipses indicating that we have omitted rules from what is given here.

(9)

```
ASSIGN_CLASS:
     <> == tq-a-70
     <1sgP - 1> == tq-o-11
     <1sgP - 12 1plF - 1> == tq-f-28
     <1sgP - 12 1plF - 1 2C - 12> == tq-zr-2
     <1sgP - 12 1plF - 1 2C - 2> == tq-zzc-1
...
```

`ASSIGN_CLASS` is a node, or location in the network, where rules are located that assign lexemes to inflectional classes. The left-hand side of each rule consists of a path, enclosed in angle brackets. The rules may differ in their degree of specificity. The first rule says that if nothing has been observed (hence the use of the empty path <>), then the lexeme will be assigned to inflectional class `tq-a-70`. The class labels are arbitrary.[16] They are morphomic (Aronoff

[16] There is a convention behind the naming, of course. All inflectional class labels start with tq, because the hierarchy was created taking into account the combination of four principal parts in section 7.3.3 ('top quad'). The alphabetic characters reflect the order in which the classes were created. For default classes, the final number represents the hypothetical size of the class, assuming that all lexemes that match with the four principal parts would be assigned to it. For small classes, the final number corresponds to the class size.

1994, Luís & Bermúdez-Otero 2016), because they are internal to the morphological system. The class labels are used to refer to nodes in the inflectional class hierarchy that define complete tone paradigms. The next rule says that if the first-person singular present has tone 1, then the lexeme will be assigned to class `tq-o-11`. Note the importance of the specificity of the information. This is illustrated by the next set of rules. If it is known that the first person singular present has tone 12 and the first plural future has tone 1, then the lexeme will be assigned to class `tq-f-28`. Knowing the tone of the second-person completive will change the class assignment to `tq-zr-2`, if the second person completive tone is 12, and to `tq-zzc-1`, if the second person completive tone is 2. This is relevant for our example *ʔien*^{12} since it has tone 12 in the first person singular. We can see here that this uses the implicational relations in the system and resorts to defaults where there may be insufficient information on which to build the assignment. For *ʔien*^{12}, if there were no observed tones, then it would be assigned to the overall default `tq-a-70`. Table 7.32 contrasts the actual tones of *ʔien*^{12} with those predicted by the default assignment rule in the absence of any further information.

The assigned tones are those defined by the default class `tq-a-70`. We can now gradually add in information, following the best combination of four cells as the guide, to see how well the assignment fares as more is known about the tone patterns of *ʔien*^{12}. If we add in information about the first-person singular for this item (Table 7.33), it will not improve things, because the first-person

Table 7.32. *Default assignment of class for* ʔien^{12} *in the absence of observed tones*

	Actual Tone	Assigned Tone	Error
1SG PRS	12	12	
1SG FUT	13	13	
1SG CPL	1	1	
1PL PRS	12	12	
1PL FUT	13	13	
1PL CPL	13	13	
2 PRS	12	12	
2 FUT	13	2	x
2 CPL	1	1	
3 PRS	12	2	x
3 FUT	1	2	x
3 CPL	1	2	x

Table 7.33. *First-person singular present*

	1SG
PRS	ʔien[12]
FUT	
CPL	

Table 7.34. *First-person plural future*

	1SG	1PL
PRS	ʔien[12]	
FUT		ʔien[13]
CPL		

singular present has already been correctly predicted by assigning the overall default class (`tq-a-70`).

Although we now know that the first-person singular present of *ʔien[12]* is 12, this will perform no better than having no observed tones. This is because the default class `tq-a-70` to which *ʔien[12]* is assigned shares a lot of the tone patterns with other tone classes that inherit from it, including the one to which *ʔien[12]* should really be assigned. Furthermore, in the assignment system there is no rule that has only the left-hand side `<1sgP - 12>`. Unless we know more about the other tones, the overall default class is the only one that can be assigned, because `<1sgP - 12>` can only match with the empty path. A similar issue arises, if we know the first person plural in addition to the first person singular (Table 7.34). This will perform no better than knowing the overall default class, again because the first-person plural future has been correctly predicted in the initial step, as shown previously in Table 7.32.

The combination of tones that *ʔien[12]* has in the first-person singular present and first-person plural future is compatible with 22 inflectional classes covering 141 lexemes. Of these inflectional classes, the default class `tq-a-70` is the largest and covers just under half of the 141 lexemes. But it is also important to note, as Table 7.32 shows, that incorrect assignment to the default class still predicts two-thirds of the paradigm correctly. A similar issue arises if we also know that the second-person completive uses tone 1 (Table 7.35).

This actually reduces the number of candidate inflectional classes down to 9 (covering 101 lexemes), but this candidate set of classes still includes the

Table 7.35. *Second-person completive*

	1SG	1PL	2
PRS	ʔien^{12}		
FUT		ʔien^{13}	
CPL			ʔien^{1}

Table 7.36. *Combination sufficient to predict the whole paradigm*

	1SG	1PL	2	3
PRS	ʔien^{12}			
FUT		ʔien^{13}		ʔien^{1}
CPL			ʔien^{1}	

default class `tq-a-70` and, given the reduction in number of lexemes covered, the default class performs even better as a predictor, accounting for two thirds of the lexemes. When we get to the combination of four cells, however, a different class is assigned, and this is sufficient to predict the rest of the paradigm (Table 7.36).

Once the four cells in Table 7.36 are known, the following assignment rule in (10) applies.

(10)

```
<1sgP - 12 1plF - 13 2C - 1 3F - 1> == tq-h-20
```

In this dataset the lexeme *ʔien*^{12} belongs to a class of sixteen lexemes to which we have given the arbitrary label `tq-h-20`. This is sufficient to predict the whole of the inflectional paradigm, as shown in Table 7.37.

So far we have seen how the cascaded default assignment works for an individual lexeme. In the next section we present the results for the whole lexicon.

7.3.5 *Results*

The results for the whole lexicon are presented in Table 7.38 giving figures for a complete combination of four cells, down through three known cells, two cells, one cell, and no information at all, scenarios that we have presented for our example verb. The total number of lexemes in the dataset is 618. The lexeme column shows how many of the 618 lexemes are correctly predicted from the given cells, and this is expressed as a percentage in the 'Lexeme Success

Table 7.37. *Assignment of class for ʔien[12] using four values*

	Actual Tone	Assigned Tone
1SG PRS	12	12
1SG FUT	13	13
1SG CPL.	1	1
1PL PRS	12	12
1PL FUT	13	13
1PL CPL	13	13
2 PRS	12	12
2 FUT	13	13
2 CPL	1	1
3 PRS	12	12
3 FUT	1	1
3 CPL	1	1

Table 7.38. *Proportion of the Tlatepuzco lexicon accounted for by a combination of default assignment and implicative relations associated with inflectional class*

Given Cells	Known Cells	Lexemes (618)	Cells (7416)	Portion of Cells	Cell Success Rate	Lexeme Success Rate
1SG PRS, 1PL FUT, 2 CPL, 3 FUT	4	487	7243	98%	97%	79%
1SG PRS, 1PL FUT, 2 CPL	3	349	6692	90%	87%	56%
1SG PRS, 1PL FUT	2	306	6441	87%	84%	50%
1SG PRS	1	183	5106	69%	66%	30%
None	0	68	2891	39%	39%	11%

Rate' Column. There are 7416 paradigm cells in total (618 x 12). The column 'Portion of Cells' indicates how many cells are correct in this scenario (including cells that are correct because they are known). The 'Cell Success Rate' is the number of correctly predicted unknown cells divided by the total number of unknown cells. For example, the cell success rate when four cells are known is

Table 7.39. *Performance of four cells (indicated by *) in predicting the rest of the paradigm*

lexemes	1SG PRS*	1SG FUT	1SG CPL	1PL PRS	1PL FUT*	1PL CPL	2 PRS	2 FUT	2 CPL*	3 PRS	3 FUT*	3 CPL	incorrect cells in class
487	correct	correct	correct	correct	correct	correct	correct	correct	correct	correct	correct	correct	0
34	correct	correct	correct	correct	correct	correct	correct	correct	correct	incorrect	correct	correct	1
20	correct	correct	correct	correct	correct	correct	correct	incorrect	correct	correct	correct	correct	1
16	correct	correct	correct	incorrect	correct	correct	correct	correct	correct	correct	correct	correct	1
14	correct	correct	correct	correct	correct	correct	correct	correct	correct	correct	correct	incorrect	1
14	correct	correct	correct	correct	correct	incorrect	correct	correct	correct	correct	correct	correct	1
10	correct	correct	correct	correct	correct	correct	incorrect	incorrect	correct	correct	correct	correct	2
3	correct	incorrect	correct	correct	correct	correct	correct	correct	correct	correct	correct	correct	1
3	correct	correct	correct	correct	correct	correct	incorrect	correct	correct	correct	correct	correct	1
3	correct	correct	correct	correct	correct	correct	incorrect	correct	correct	correct	correct	incorrect	2
3	correct	correct	correct	correct	correct	correct	incorrect	incorrect	correct	incorrect	correct	incorrect	4
2	correct	correct	correct	correct	correct	incorrect	correct	incorrect	correct	correct	correct	correct	2
2	correct	correct	correct	correct	correct	correct	correct	correct	correct	incorrect	correct	incorrect	2
1	correct	correct	incorrect	correct	correct	correct	correct	correct	correct	correct	correct	correct	1
1	correct	correct	correct	incorrect	correct	correct	correct	correct	correct	incorrect	correct	correct	2
1	correct	correct	correct	correct	correct	incorrect	correct	correct	correct	incorrect	correct	incorrect	3
1	correct	correct	correct	correct	correct	correct	incorrect	incorrect	correct	correct	correct	incorrect	3
1	correct	correct	correct	incorrect	correct	incorrect	correct	correct	correct	incorrect	correct	incorrect	4
1	correct	correct	correct	incorrect	correct	incorrect	correct	incorrect	correct	incorrect	correct	incorrect	5
1	correct	correct	correct	incorrect	correct	correct	incorrect	incorrect	correct	incorrect	correct	incorrect	5
% correct	100.0%	99.5%	99.8%	96.7%	100.0%	96.9%	96.6%	93.8%	100.0%	92.8%	100.0%	95.6%	
% correct	99.4%			94%			92.8%			89.6%			

97 percent. In this scenario, the total number of known cells is 2472 (618 x 4). The total number of unknown cells is 4944 (7416–2472). The number of correctly predicted unknown cells is 4771 (7243–2472). This gives the cell success rate of 97 percent (4771/4944). The figures for the other combinations are arrived at in the same way, although the number of unknown cells will increase when there is a smaller number of known cells.

There are 487 lexemes whose paradigms are correctly predicted. In Table 7.39 we see that there are 592 (95.8%) lexemes for which one cell or less is incorrectly predicted. That means 105 lexemes that contain one override in their lexical entries. Of the remaining 26 lexemes, 18 require two overrides, 2 lexemes require 3 overrides, 4 lexemes require 4 overrides, and 2 require 5 overrides. No lexeme needs to specify more than 5 paradigm cells for the 12-cell paradigm dealt with here.

7.4 Conclusion

The Tlatepuzco Chinantec verbal lexicon presents an intricate interplay of lexical stipulation and grammatical inference. In this chapter, we have associated lexical stipulation with emergent complexity, uncertainty about exponence, whereas the morphological grammar is associated with the opposite, organization, in which there is certainty about exponence. We argued that it is the compromise between the two extremes that makes the Tlatepuzco Chinantec system, in one sense, complex, and we have used the concept of central system complexity to characterize this. Abstract paradigm types were introduced to illustrate how different types of complexity relate to one another. Real language systems are, of course, mixtures of these types. We saw how the Tlatepuzco Chinantec verbal system also contains, at a higher scale called the inflectional series, hierarchical structure, something that we associate with high central system complexity. The systematic nature of inflectional series underscores the need to observe the morphological system with a mind to different scales of structure from individual paradigm cells, through the inflectional series we observed for Tlatepuzco Chinantec up to the whole paradigm. This is our motivation for adopting a composite analysis of the Tlatepuzco Chinantec system so as to determine how it works at different morphological scales. The first analysis we presented showed how the mid-scale inflectional series structures created reliable associations between person and tone patterns. We demonstrated that there are strong implicational relations between the series. We then went on to consider individual paradigm cells, as used in a principal part analysis, to show how inference at the level of the paradigm cell and the inflectional series can work together. We also noted that the best combination of paradigm cells used for the principal part analysis exhibits the property of value maximization – namely, providing information about all the relevant values. We then introduced a model that is able to work with varying degrees of principal-part information, from nothing to four paradigm cells. Defaults are crucial for this to work. As such, they are an ideal means for navigating between the different morphological scales, from information based on individual cells all the way to the whole paradigm.

8 *Morphological Complexity and Morphological Autonomy*

Inflection is the encounter of the lexicon with functional categories – that is, syntactic and semantic features that cross-cut lexical distinctions – resulting in the hybrid we know as inflectional morphology. Although it is perfectly possible for the two domains to maintain their segregation even within the narrow confines of a single word form (recall the Aymara noun paradigms from the Introduction), the preceding chapters have given ample evidence of their intensive interaction. The fruit of this is complexity – *morphological complexity* – by which we mean those additional categories, both lexical and paradigmatic, that often emerge in inflectional systems. To some extent, this is a problem of our own making – our meaning any of us whose task it is to describe language. Would it not be easier to assume that we can reliably read the functional categories directly off the morphological forms? That is possible with Aymara nouns, but with many other languages the result would be an utter mess, resulting in the multiplication of the rules of syntax needed to account for agreement and government across different sets of lexemes. Imagine what a description of Tlatepuzco Chinantec syntax would look like if it had to account for all the forms described in Chapter 7! If a parsimonious account of grammar is something to aim for, morphological complexity is an unavoidable by-product. But although this gives us a general strategy for isolating morphological complexity, it hardly gives us a foolproof formula for an analysis. The division between motivated and arbitrary morphological distinctions depends on an all-or-nothing conception of the assignment of morphological form: either it is predictable on the basis of some non-morphological property or it is not. But, in practice, the division may well be fluid. As we have seen, particularly in Chapters 4 and 5, motivation is sometimes approximate; for example, semantic or phonological properties may partly predict morphological behaviour without our quite being able to say that they assign it. This muddies the waters, making it hard to gauge the precise dimensions of morphological complexity in any given instance, let alone across the whole array of languages whose descriptions we have access to. The typological frameworks we have offered in Chapter 6, and the measures derived from them explored in Chapter 7, are of course based on a forced choice, a willful suppression of nuances and ambiguities. But this is in the nature of the typological enterprise, especially since the grammatical models we all work with are expressed in categorical terms.

Given the contingent nature of its definition, the dimensions of morphological complexity will vary according to the sorts of analytical decisions the individual investigator makes. Having then isolated it – it will always be present to some degree, we would argue – where in the larger scheme of language does it fit? For some the answer is obvious: in a separate module of grammar dedicated to the structure of and relationships among word forms. These and allied phenomena are taken as evidence of an autonomous morphological component of language. For others this smacks of redundancy: phonology, syntax, semantics, and the lexicon are needed for any linguistic description, and by reference just to these we can construct many inflectional paradigms. As for the residue not covered by transparent mappings among components, these can be handled by more involved interface rules, or indeed at some point allowed simply to be exceptions, accidental by-products of rules gone awry. It should be no secret to anyone who has at least skimmed the preceding chapters that our sympathies lie with the former view. But although this has provided a powerful motivation to look into the question, the typological facts are there in any case, no matter where they fit into one's philosophy of language. And the facts are that inflectional paradigms have the properties that they do, and are learnt as such. If our model of lexical, morphosyntactic, and morphosemantic properties were based solely on these objects, it would no doubt look very different from what we normally entertain. It may make perfect sense that we do not do this, that the morphological facts are brought into line with the categories that make sense for the other linguistic components, but we should recall that this is a choice and not an inevitability.

If we do entertain the notion of autonomous morphology, it is still by no means obvious what it should look like. In practice, there are two diagnostics. The first, of course, is absence of any correspondence with categories found elsewhere in the grammar. This is not to say that where there are such correspondences this necessarily means that morphology must play second fiddle to some other grammatical component. For example, if some set of inflectional distinctions corresponds to a set of semantic distinctions, must it be that morphology is parasitic of semantics, or could the semantic categories be supported and propagated by the morphological distinctions? But in such situations, morphology has at least a dancing partner, so if not leading, it is, at any rate, not autonomous. With those structures that remain, the question typically posed is: are they systematic, or are they accidental? Translated into a cognitive metaphor, if not a model, do speakers 'know' inflection classes, be they defined by allomorphy or the varying partitions of the paradigmatic space that we have termed distributional classes? Or are these in some way ephemeral, a series of isolated facts with no superstructure to maintain them? We can take various pieces of evidence as indications of their systematicity, in particular the propagation of patterns across generations, also independent measures such as brain imaging, and, of course, the very fact of their acquisition in the first place. But we would also suggest that one should not be overly seduced by a priori notions of systematicity as an objective and reliable measure of the ontological status of morphology. Because this is the way language behaves or can behave.

Appendix

This appendix lists the inflectional series used in Palancar's (2014) classification of the tone patterns discussed in Chapter 7. The plus and minus marks are instructions to add or remove ballistic stress, where relevant. Although it plays a role in distinguishing between a small number of the series, this operation is not relevant for our analysis in Chapter 7. The series indicated by capital or lowercase letters are *not* used to realize the third person. As can be seen, inflectional series involving roman numerals do not distinguish TAM. Arabic series differentiate at least two TAM values. Inflectional series a, b, c (lower case letters) are restricted to use within the first person.

Series	PRS	CPL	FUT
I	+2	+2	+2
II	12	12	12
III	1	1	1
IV	13	13	13
V	-3	-3	-3
1	12	1	1
2	-2	-2	1
3	12	+2	1
4	2	1	1
5	12	1	13
6	2	1	2
7	-3	-3	1
8	12	1	2
9	12	1	12
10	2	1	13
A	2	-3	+3
B	2	2	+3
C	+2	+1	+3
D	12	13	13
E	12	12	13
F	2	-12	+3
G	12	-3	13
a	+2	+2	+3
b	2	+3	+3
c	2	13	13

References

Ackerman, Farrell & Robert Malouf. 2013. Morphological organization: The low conditional entropy conjecture. *Language* 89. 429–464.

Albright, Adam. 2002. Islands of reliability for regular morphology: Evidence from Italian. *Language* 78. 684–709.

Anderson, Stephen R. 1992. *A-Morphous Morphology*. Cambridge: Cambridge University Press.

2008. Phonologically conditioned allomorphy in the morphology of Surmiran (Rumantsch). *Word Structure* 1. 109–134.

Aoki, Haruo. 1994. *Nez Perce Dictionary*. (University of California Publications in Linguistics 122). Berkeley: University of California Press.

Arensen, Jon. 1982. *Murle Grammar* (Occasional papers in the study of Sudanese languages 2). Juba: University of Juba and SIL.

Aronoff, Mark. 1994. *Morphology by Itself: Stems and Inflectional Classes* (Linguistic Inquiry Monograph 22). Cambridge, MA: MIT Press.

Austin, Peter. 1981/2013. *A Grammar of Diyari, South Australia*. Cambridge: Cambridge University Press. 2nd edn., version 2.5, 2013, School of Oriental and African Studies, London. [Page references to the 2013 edition.]

Baayen, R. Harald. 2008. *Analyzing Linguistic Data. A Practical Introduction to Statistics Using R*. Cambridge: Cambridge University Press.

Baerman, Matthew. 2001. Unnatural classes in morphological change. *Russian Linguistics* 25. 281–284.

2012. Paradigmatic chaos in Nuer. *Language* 88(3). 467–494.

2014. Covert systematicity in a distributionally complex system. *Journal of Linguistics* 50. 1–47.

2016. Seri verb classes: morphosyntactic motivation and morphological autonomy. *Language* 92(4). 792–823.

Baerman, Matthew, Dunstan Brown, & Greville G. Corbett. 2005. *The Syntax-Morphology Interface: A Study of Syncretism*. Cambridge: Cambridge University Press.

Baerman, Matthew & Greville G. Corbett. 2012. Stem alternations and multiple exponence. *Word Structure* 5.1. 52–68.

Baerman, Matthew, Greville G. Corbett, & Dunstan Brown (eds). 2010. *Defective Paradigms: Missing Forms and What They Tell Us* (Proceedings of the British Academy, 163). Oxford: British Academy and Oxford University Press.

Baerman, Matthew, Greville Corbett, Dunstan Brown, & Andrew Hippisley (eds). 2007. *Deponency and Morphological Mismatches* (Proceedings of the British Academy, 145). Oxford: British Academy and Oxford University Press.

Baerman, Matthew & Enrique L. Palancar. 2014. The organization of Chinantec tone paradigms. *Carnets de Grammaire 22: Proceedings of Les Décembrettes, 8th International Conference on Morphology*, 46–59. Toulouse: CNRS & Université Toulouse – Jean Jaurès.

Blevins, James P. 2003. Stems and paradigms. *Language* 79 (4). 737–767.

2004. Inflection classes and economy. In: Gereon Müller, Lutz Gunkel & Gisela Zifonun (eds) *Explorations in Nominal Inflection*, 41–85. Berlin: Mouton de Gruyter.

2006. Word-based morphology. *Journal of Linguistics* 42. 531–573.

Blevins, James P., Farrell Ackerman, Robert Malouf, & Michael Ramscar. 2016. Morphology as an adaptive discriminative system. In: Daniel Siddiqi & Heidi Harley (eds) *Morphological Metatheory*, 271–301. Amsterdam: John Benjamins.

Bobaljik, Jonathan. 2000. The ins and outs of contextual allomorphy. In: Kleanthes K. Grohmann & Caro Struijke (eds) *University of Maryland Working Papers in Linguistics* 10. 35–71.

Bond, Oliver. 2013. A base for canonical negation. In: Dunstan Brown, Marina Chumakina & Greville G. Corbett (eds) *Canonical Morphology and Syntax*, 20–47. Oxford: Oxford University Press.

Briley, David. 1997. Four grammatical marking systems in Bauzi. In: Karl J. Franklin (ed.) *Papers in Papuan Linguistics*, 2, 1–131. Canberra: Research School of Pacific and Asian Studies, Australian National University.

Brown, Dunstan. 1998. From the general to the exceptional: A network morphology account of Russian nominal inflection. PhD thesis, University of Surrey.

2007. Peripheral functions and overdifferentiation: The Russian second locative. *Russian Linguistics* 31. 61–76.

Brown, Dunstan & Roger Evans. 2012. Morphological complexity and unsupervised learning: Validating Russian inflectional classes using high frequency data. In: Ferenc Kiefer, Mária Ladányi & Péter Siptár (eds) *Current Issues in Morphological Theory: (Ir)regularity, Analogy and Frequency*. Selected Papers from the 14th International Morphology Meeting, Budapest, 13–16 May 2010, 135–162. Amsterdam: John Benjamins.

Brown, Dunstan & Andrew Hippisley. 2012. *Network Morphology: A Defaults-based Theory of Word Structure*. Cambridge: Cambridge University Press.

Browne, Wayles. 1993. Serbo-Croat. In: Bernard Comrie & Greville G. Corbett (eds) *The Slavonic Languages*, 306–387. London: Routledge.

Bugarski, Ranko. 2012. Language, identity and borders in the former Serbo-Croatian area. *Journal of Multilingual and Multicultural Development* 33. 219–235.

Bunn, Gordon. 1974. *Golin Grammar*. Ukarumpa: SIL. Online: www.sil.org/pacific/png/abstract.asp?id=10339.

Bye, Patrik. 2015. The nature of allomorphy and exceptionality: Evidence from Burushaski plurals. In: Eulàlia Bonet, Maria-Rosa Lloret & Joan Mascaró (eds) *Understanding Allomorphy: Perspectives from Optimality Theory*, 107–176. London: Equinox.

Caballero, Gabriela. 2011. Morphologically conditioned stress assignment in Choguita Rarámuri (Tarahumara). *Linguistics* 49(4). 749–790.

Campbell, Eric. 2011. Zenzontepec Chatino aspect morphology and Zapotecan verb classes. *International Journal of American Linguistics* 77(2). 219–246.

Carleton, Troi & Rachelle Waksler. 2000. Pronominal markers in Zenzontepec Chatino. *International Journal of American Linguistics* 66(3). 383–397.

Carstairs, Andrew. 1983. Paradigm economy. *Journal of Linguistics* 19. 115–128.

1987. *Allomorphy in Inflexion*. London: Croom Helm.

1988. Some implications of phonologically conditioned suppletion. In: Geert Booij & Jaap van Marle (eds) *Yearbook of Morphology 1988*, 67–94. Dordrecht: Foris.

1990. Phonologically conditioned suppletion. In: Wolfgang U. Dressler, Hans C. Luschützky, Oskar E. Pfeiffer, & John R. Rennison (eds) *Contemporary Morphology*, 17–23. Berlin: Mouton de Gruyter.

Carstairs-McCarthy, Andrew. 1994. Inflection classes, gender, and the principle of contrast. *Language* 70. 737–788.

2010. *The Evolution of Morphology*. Oxford: Oxford University Press.

Cerrón-Palomino, Rodolfo. 2006. *El chipaya o la lengua de los hombres del agua*. Lima: Pontificia Univ. Católica del Perú.

Chumakina, Marina & Greville G. Corbett. 2015. Gender-number marking in Archi: Small is complex. In: Matthew Baerman, Dunstan Brown, & Greville G. Corbett (eds) *Understanding and Measuring Morphological Complexity*, 93–116. Oxford: Oxford University Press.

Chumakina, Marina, Dunstan Brown, Greville G. Corbett, & Harley Quilliam. 2007. *Archi: A Dictionary of the Archi villages, Southern Daghestan, Caucasus*. Online: www.smg.surrey.ac.uk/archi/linguists/.

Ciucci, Luca. 2007. Indagini sulla morfologia verbale nella lingua ayoreo. *Quaderni del Laboratorio di Linguistica* 7. Pisa: Scuola Normale Superiore.

Ciucci, Luca & Pier Marco Bertinetto. 2015. A diachronic view of Zamucoan verb inflection. *Folia Linguistica* 36. 19–87.

Coler, Matt. 2015. Aymara inflection. In: Matthew Baerman (ed.) *The Oxford Handbook of Inflection*, 521–542. Oxford: Oxford University Press.

Corbett, Greville G. 1982. Gender in Russian: An account of gender specification and its relationship to declension. *Russian Linguistics* 6. 197–232.

1991. *Gender*. Cambridge: Cambridge University Press.

2000. *Number*. Cambridge: Cambridge University Press.

2006. *Agreement*. Cambridge: Cambridge University Press.

2009a. Canonical inflectional classes. In: Fabio Montermini, Gilles Boyé & Jesse Tseng (eds) *Selected Proceedings of the 6th Décembrettes: Morphology in Bordeaux*, 1–11. Somerville, MA: Cascadilla Proceedings Project. Online: www.lingref.com/cpp/decemb/6/abstract2231.html.

2009b. Morphology-free syntax: Two potential counter-examples from Serbo-Croat. In: Steven Franks, Vrinda Chidambaram, & Brian Joseph (eds) *A Linguist's Linguist: Studies in South Slavic Linguistics in Honor of E. Wayles Browne*, 149–166. Bloomington, IN: Slavica.

2010. Classic problems at the syntax-morphology interface: Whose are they? In: Stefan Müller (ed.) *Proceedings of the HPSG 2010 Conference*, 255–268. Paris:

Université Paris Diderot, Paris 7. CSLI Publications. Online: http://cslipublications .stanford.edu/HPSG/2010/toc.shtml.

2012. *Features*. Cambridge: Cambridge University Press.

2013. Paradigm conventions. Paper presented at the 46th Annual Meeting of the Societas Linguistica Europaea, Split, 18–21 September 2013. Online: www .academia.edu/9055930/Paradigm_conventions.

2014. Lexicalization and paradigmatic structure: Key instances in Slavonic. In: Motoki Nomachi, Andrii Danylenko, & Predrag Piper (eds) *Grammaticalization and Lexicalization in the Slavic Languages: Proceedings from the 36th meeting of the Commission on the Grammatical Structure of the Slavic Languages of the International Committee of Slavists* (Die Welt der Slaven: Sammelbände – Sborniki, 55), 266–274. Munich: Otto Sagner.

2015. Morphosyntactic complexity: A typology of lexical splits. *Language* 91. 145–193.

2016. Morphomic splits. In: Ana Luís & Ricardo Bermúdez-Otero (eds) *The Morphome Debate: Diagnosing and analysing morphomic patterns*, 64–88. Oxford: Oxford University Press.

Corbett, Greville G. & Matthew Baerman. 2006. Prolegomena to a typology of morphological features. *Morphology* 16. 231–246.

Corbett, Greville G. & Wayles Browne. 2009. Serbo-Croat: Bosnian, Croatian, Montenegrin, Serbian (=Chapter 18). In: Bernard Comrie (ed.) *The World's Major Languages*, 2nd edn., 330–346. London: Routledge. [Revised and updated version of 1987 chapter.]

Corbett, Greville G. & Norman M. Fraser. 1993. Network Morphology: A DATR account of Russian inflectional morphology. *Journal of Linguistics* 29. 113–42. [Reprinted 2003 in: Francis X. Katamba (ed.) *Morphology: Critical Concepts in Linguistics: VI: Morphology: Its Place in the Wider Context*, 364–396. (London: Routledge.)]

1997. Vyčislitel´naja lingvistika i tipologija. *Vestnik MGU: Serija 9: Filologija* no. 2, 122–140.

2000. Gender assignment: A typology and a model. In: Gunter Senft (ed.) *Systems of Nominal Classification* (Language, Culture and Cognition 4), 293–325. Cambridge: Cambridge University Press.

Crowell, Thomas Harris. 1979. A grammar of Bororo. PhD thesis, Cornell University.

Crysmann, Berthold & Olivier Bonami. 2016. Variable morphotactics in information-based morphology. *Journal of Linguistics* 52. 311–374.

Davidson, Matthew. 2002. Studies in southern Wakashan (Nootkan) grammar. PhD thesis, State University of New York at Buffalo. Online: http://wings.buffalo.edu/linguistics/people/students/dissertations/davidson.PDF.

Davies, John. 1981. *Kobon*. Amsterdam: North Holland Publishing Company.

Davis, Irvine. 1964. The Language of Santa Ana Pueblo. Smithsonian Institution, Bureau of American Ethnology, Bulletin 191, Anthropological Papers No. 69. 53–190. Washington, DC, US Government Printing Office.

Dimmendaal, Gerrit J. 2000. Number marking and noun categorization in Nilo-Saharan languages. *Anthropological Linguistics* 42. 214–261.

Donohue, Mark. 2001. Animacy, class and gender in Burmeso. In: Andrew Pawley, Malcolm Ross, & Darrell Tryon (eds) *The Boy from Bundaberg: Studies in*

Melanesian Linguistics in Honour of Tom Dutton (Pacific linguistics 514), 97–115. Canberra: Pacific Linguistics.

Dressler, Wolfgang U. 2003. Degrees of grammatical productivity in inflectional morphology. *Rivista di Linguistica* 15. 31–62.

Dressler, Wolfgang U., Marianne Kilani-Schoch, Natalia Gagarina, Lina Pestal, & Markus Pöchtrager. 2006. On the typology of inflection class systems. *Folia Linguistica* 40. 51–74.

Dum-Tragut, Jasmine. 2009. *Armenian: Modern Eastern Armenian*. Amsterdam: John Benjamins.

Dybo, Vladimir A. 2000. *Morfonologizovannye paradigmatičeskie akcentnye sistemy: Tipologija i genezis*. Moscow: Jazyki russkoj kul´tury.

Embick, David. 2003. Locality, listedness, and morphological information. *Studia Linguistica* 57. 143–169.

Embick, David & Rolf Noyer. 2007. Distributed morphology and the syntax—morphology interface. In: Gillian Ramchand & Charles Reiss (eds) *The Oxford Handbook of Linguistic Interfaces*, 289–324. Oxford: Oxford University Press.

Erelt, Mati, Tiiu Erelt, & Kristiina Ross. 1997. *Eesti keele käsiraamat*. Tallinn: Eesti Keele Sihtasutus. Online: www.eki.ee/books/ekkr/.

Evans, Nicholas, Jutta Besold, Hywel Stoakes, & Alan Lee (eds). 2005. *Materials on Golin: Grammar, texts and dictionary*. Melbourne: Deptartment of Linguistics and Applied Linguistics, The University of Melbourne.

Evans, Roger & Gerald Gazdar. 1996. DATR: A language for lexical knowledge representation. *Computational Linguistics* 22. 167–216.

Fähnrich, Heinz. 2012. *Die georgische Sprache*. Leiden: Brill.

Feist, Timothy & Enrique L. Palancar. 2015. Oto-Manguean Inflectional Class Database: Tlatepuzco Chinantec. University of Surrey. Online: http://dx.doi.org/10.15126/SMG.28/1.01.

Feldstein, Ronald. 2001. *A Concise Polish Grammar*. Durham and Chapel Hill: Slavic and East European Language Resource Center.

Foley, William A. 1986. *The Papuan Languages of New Guinea*. Cambridge: Cambridge University Press.

Franks, Steven L. 1990. Vowel-zero alternations and syllable-counting morphology. *Indiana Slavic Studies* 5. 79–96.

Gardani, Francesco. 2013. *Dynamics of Morphological Productivity: The Evolution of Noun Classes from Latin to Italian*. Leiden: Brill.

Gershenson, Carlos & Nelson Fernández. 2012. Complexity and information: Measuring emergence, self-organization, and homeostasis at multiple scales. *Complexity* 18. 29–44.

Gorbov, Andrej A. 2014. Čislovye paradigmy abstraknyx suščestvitel´nyx v russkom jazyke XX veka: tendencii razvitija i vlijanie anglijskogo jazyka. [Number paradigms of abstract nouns in twentieth-century Russian: Development trends and the influence of English] *Russian Linguistics* 38. 23–46.

Hernández-Green, Néstor. 2015. *Morfosintaxis del otomí de Acazulco*. PhD thesis, CIESAS, Mexico.

Himmelmann, Nikolaus P. 2014. Asymmetries in the prosodic phrasing of function words: Another look at the suffixing preference. *Language* 90(4). 927–960.

Huntley, David. 1980. The evolution of genitive-accusative animate and personal nouns in Slavic dialects. In: Jacek Fisiak (ed.) *Historical Morphology* (Trends in Linguistics: Studies and Monographs 17), 189–212. The Hague: Mouton.

Isačenko, Alexander V. 1962. *Die russische Sprache der Gegenwart* (Teil I. *Formenlehre)*. Halle: Max Niemeyer.

Iva, Sulev. 2007. Võru kirjakeele sõnamuutmissüsteem. PhD thesis, University of Tartu.

Ivić, Pavle. 1972. Sistema padežnyx okončanij suščestvitel´nyx v serboxorvatskom literaturnom jazyke. In: F. P. Filin (ed.) *Russkoe i slavjanskoe jazykoznanie: K 70-letiju člena-korrespondenta AN SSSR R. I .Avanesova*, 106–121. Moscow: Nauka. [Republished as Sistem padežnih nastavaka imenica u srpskohrvatskom književnom jeziku, *in* Pavle Ivić (1990) *O jeziku nekadašnjem i sadašnjem*, 286–309. Belgrade: BIGS-Jedinstvo.]

Janda, Laura A. & Charles E. Townsend. 2002. *Czech*. SEELRC. Online: www.seelrc.org:8080/grammar/mainframe.jsp?nLanguageID=2.

Kemmer, Susanne. 1993. *The Middle Voice*. Amsterdam-Philadelphia: John Benjamins.

Kibrik, Aleksandr E. 1977. *Opyt strukturnogo opisanija arčinskogo jazyka: II: Taksonomičeskaja grammatika*. (Publikacii otdelenija strukturnoj i prikladnoj lingvistiki 12). Moscow: Izdatel´stvo Moskovskogo universiteta.

2003. Nominal inflection galore: Daghestanian, with side glances at Europe and the world. In: Frans Plank (ed.) *Noun Phrase Structure in the Languages of Europe*, 37–112. Berlin: Mouton de Gruyter.

Kibrik, Aleksandr E., S. V. Kodzasov, I. P. Olovjannikova, & D. S. Samedov. 1977. *Opyt strukturnogo opisanija arčinskogo jazyka: I: Leksika, fonetika*. (Publikacii otdelenija strukturnoj i prikladnoj lingvistiki, 11). Moscow: Izdatel´stvo Moskovskogo universiteta.

Klaić, Bratoljub. 1953. O dugoj množini trosložnih i višesložnih imenica. *Jezik* (Zagreb) 2, no. 3. 74–77.

Kulikov, Leonid. 2004. Review of *Number*. *Southwest Journal of Linguistics* 23. 124–129.

Laskowski, Roman. 1990. The structure of the inflectional paradigm. *ScandoSlavica* 36. 149–154.

Ljaševskaja, Ol´ga N. 2004. *Semantika russkogo čisla*. Moscow: Jazyki Slavjanskoj Kul´tury.

Loporcaro, Michele. 2007. On triple auxiliation in Romance. *Linguistics* 45, 1, 173–222.

Luís, Ana & Ricardo Bermúdez-Otero (eds). 2016. *The Morphome Debate*. Oxford: Oxford University Press.

Luiten, Tyler V. 2011. Old High German nominal morphology revisited: A fresh look at old paradigms. PhD dissertation, University of Wisconsin-Madison.

Maiden, Martin. 1992. Irregularity as a determinant of morphological change. *Journal of Linguistics* 28. 285–312.

2005. Morphological autonomy and diachrony. *Yearbook of Morphology* 2004, 137–75.

2011. Morphological persistence. In: Martin Maiden, John Charles Smith & Adam Ledgeway (eds) *The Cambridge History of the Romance Languages* (Volume 1: *Structures*), 155–215. Cambridge: CUP.

2013. The Latin ‘third stem’ and its Romance descendants. *Diachronica* 30. 492–530.

Marlett, Stephen A. 1981. The structure of Seri. PhD dissertation, University of California at San Diego.

2016. *Cmiique Iitom (The Seri language)*. Ms., University of North Dakota. Online: http://arts-sciences.und.edu/summer-institute-of-linguistics/faculty/marlett-steve/marlett-seri-grammar-latest-draft.pdf.

Mathiassen, Terje. 1997. *A Short Grammar of Latvian*. Columbus, Ohio: Slavica.

McElhanon, Kenneth A. 1972. *Selepet Grammar; Part I: From Root to Phrase* (Pacific Linguistics, Series B, No. 21). Canberra: Australian National University.

Mel′čuk, Igor′ A. 1985. *Poverxnostnyj sintaksis russkix čislovyx vyraženij* (Wiener Slawistischer Almanach: Sonderband 16). Vienna: Institut für Slawistik der Universität Wien.

Merrifield, William R. & Alfred E. Anderson. 2007. *Diccionario Chinanteco de la diáspora del pueblo antiguo de San Pedro Tlatepuzco, Oaxaca*. [2nd edn.]. Mexico DF: Summer Institute of Linguistics.

Metzger, Ronald G. 2000. *Marĩ yaye mena carapana, yaia yaye mena español macãrĩcã tuti = Carapana-español, diccionario de 1000 palabras*. Bogotá: Editorial Alberta Lleras Camargo.

Miestamo, Matti. 2006. On the feasibility of complexity metrics. In: Krista Kerge & Maria-Maren Sepper (eds), *FinEst Linguistics: Proceedings of the Annual Finnish and Estonian Conference of Linguistics, Tallinn, 6–7 May 2004*, 11–26. Tallinn: Tallinn University Department of Estonian.

2008. Grammatical complexity in a cross-linguistic perspective. In: Matti Miestamo, Kaius Sinnemäki, & Fred Karlsson (eds), *Language Complexity: Typology, Contact, Change*, 23–41. Amsterdam and Philadelphia: John Benjamins.

Mikaelian, Irina. 2013. Cardinal numeral constructions and the category of animacy in Russian. *Russian Linguistics* 37. 71–90. DOI: 10.1007/s11185-012-9105-3.

Milanović, Branislav. 1949. *Mrazovi: mrazevi; nosovi: nosevi i sl. Naš jezik* (Belgrade) I/1–2. 43–53.

Miller, Amy. 2001. *A Grammar of Jamul Tiipay*. Berlin: Mouton de Gruyter.

Miller, Philip, Geoffrey K. Pullum, & Arnold M. Zwicky. 1997. The principle of phonology-free syntax: Four apparent counterexamples in French. *Journal of Linguistics* 33.1. 67–90.

Moser, Mary B. & Stephen A. Marlett. 2010. *Comcaac quih yaza quih hant ihiip hac = Diccionario seri-español-inglés*. Colección Bicentenario. Mexico City and Sonora: Plaza y Valdés Editores and Universidad de Sonora. Online: www.sil.org/resources/archives/42821.

Müller, Gereon. 2007. Notes on paradigm economy. *Morphology* 17. 1–38.

Naba, Jean-Claude. 1994. *Le Gulmancema: Essai de systèmatisation*. Köln: Rüdiger Köppe.

Nichols, Johanna. 2011. Ingush Grammar. (University of California Publications Series.) Berkeley: University of California Press.

2015. Types of spread zones: Open and closed, horizontal and vertical. In: Rik De Busser & Randy J. LaPolla (eds) *Language, Structure and Environment: Social, Cultural, and Natural Factors*, 261–86. Amsterdam: John Benjamins.

Nikolić, Miroslav B. 2013. Imenice koje u srpskom književnom jeziku proširuju osnovu morfemom *-ov-* u množini. *Južnoslovenski filolog* 69. 277–318.

Olander, Thomas. 2009. *Balto-Slavic Accentual Mobility*. Berlin: Mouton de Gruyter.
Orlandi, Roberto. 1963. Il plurale breve e lungo in serbo-croato. *Ricerche slavistiche* 11.3–33.
Orr, Robert. 1996. Again the *ŭ-stems in common Slavic. *Journal of Slavic Linguistics* 4.312–343.
Ouoba, Bendi Benoît. 1982. Description systèmatique du gulmancema. PhD Thesis, Institut National des Langues et Civilisations Orientales (INALCO), Université de la Sorbonne Nouvelle.
Palancar, Enrique L. 2012. The conjugation classes of Tilapa Otomi: An approach from canonical typology. *Linguistics* 50.4. 783–832.
2014. Revisiting the complexity of the Chinantec verb conjugation classes. In: Jean-Léo Léonard & Alain Kihm (eds) *Patterns in Mesoamerican Morphology*, 77–102. Paris: Michel Houdiard.
Palancar, Enrique L. & Matthew Baerman. 2014. Reaching out beyond the word form: Periphrasis and inflectional classes. Paper at the 16th International Morphology Meeting, Budapest, 31 May.
Paster, Mary. 2015. Phonologically conditioned suppletive allomorphy: Cross-linguistic results and theoretical consequences. In: Eulàlia Bonet, Maria-Rosa Lloret & Joan Mascaró (eds) *Understanding Allomorphy: Perspectives from Optimality Theory*, 218–253. London: Equinox.
Pelletier, Francis Jeffry. 2012. Lexical nouns are both +mass and +count, but they are neither +mass nor +count. In: Diane Massam (ed.) *Count and Mass Across Languages* (Oxford Studies in Theoretical Linguistics 42), 9–26. Oxford: Oxford University Press.
Pešikan, Mitar. 1956. O umetku *-ov- (-ev-)* u množini imenica prve vrste. *Naš jezik* 8. 270–275.
Pitkin, Harvey. 1984. *Wintu Grammar*. Berkeley: University of California Press.
Plungian [Plungjan], Vladimir A. 2002. K semantike russkogo lokativa. *Semiotika i informatika* 37. 229–254.
Polivanova, A. K. 1983. Vybor čislovyx form suščestvitel´nogo v russkom jazyke. In: V. P. Grigor´ev (ed.) *Problemy strukturnoj lingvistiki*, 130–145. Moscow: Nauka.
Rätsep, Huno. 1982. *Eesti keele ajalooline morfoloogia* (Vol. 1, 2nd edn.). Tartu: Tartu Riiklik Ülikool.
Reh, Mechthild. 1985. *Die Krongo-Sprache (Nìinò Mó-Dì): Beschreibung, Texte, Wörterverzeichnis*. Berlin: Dietrich Reimer Verlag.
Rich, Rolland. 1999. *Diccionario Arabela-Castellano*. (Serie Lingüística Peruana, 49.) Lima, Peru: Instituto Lingüístico de Verano.
Robinson, Stewart. 2011. Split intransitivity in Rotokas, a Papuan language of Bougainville. PhD thesis, Radboud University.
Rodrigues, Aryon Dall'Igna. 1993. Uma hipótese sobre a flexão de pessoa em Boróro. *Anais da 45a Reunião Anual da SBPC, Recife*, p. 50. Online: http://biblio.etnolinguistica.org/rodrigues-1993-bororo.
Romero-Mendez, Rodrigo. 2008. A reference grammar of Ayutla Mixe. PhD thesis, State University of New York at Buffalo.
Rothstein, Robert A. 1993. Polish. In: Bernard Comrie & Greville G. Corbett (eds) *The Slavonic Languages*, 686–758. London: Routledge.

Rupp, James & Nadine Rupp. 1996. *Diccionario chinanteco de San Juan Lealao, Oaxaca*. Tucson: Summer Institute of Linguistics.

Saeed, John I. 1999. *Somali*. Amsterdam: John Benjamins.

Sagna, Serge. 2012. Physical properties and culture-specific factors as principles of semantic categorisation of the Gújjolaay Eegimaa noun class system. *Cognitive Linguistics* 23.1. 129–163.

Sakhno, Sergueï. 2011. Pourquoi deux génitifs et deux locatifs en russe pour certains substantifs? Etat actuel des paradigmes et aspects diachroniques. In: Michèle Fruyt, Michel Mazoyer & Dennis Pardee (eds) *Grammatical Case in the Languages of the Middle East and Europe: Acts of the International Colloquium Variations, Concurrence et Evolution des Cas dans Divers Domaines Linguistiques, Paris, 2–4 April, 2007*, 359–372. Chicago: University of Chicago Press.

Samardžija, Marko. 1988. Duga i kratka množina u hrvatskom književnom jeziku. *Jezik* (Zagreb) XXXV/5. 129–136.

Sapir, Edward. 1922. The Takelma language of southwestern Oregon. In: Franz Boas (ed.) *Handbook of American Indian Languages* [vol. 2], pp. 1–296. Washington, DC: US Government Printing Office.

Schenker, Alexander M. 1993. Proto-Slavonic. In: Bernard Comrie & Greville G. Corbett (eds) *The Slavonic Languages*, 60–121. London: Routledge.

Seiler, Walter. 1985. *Imonda: A Papuan Language* (Pacific Linguistics, Series B, No. 93). Canberra: Australian National University.

Short, David. 1993a. Czech. In: Bernard Comrie & Greville G. Corbett (eds) *The Slavonic Languages*, 455–532. London: Routledge.

1993b. Slovak. In: Bernard Comrie & Greville G. Corbett (eds) *The Slavonic Languages*, 533–592. London: Routledge.

Siewierska, Anna. 2004. *Person*. Cambridge: Cambridge University Press.

Sims, Andrea. 2015. *Inflectional Defectiveness*. Cambridge: Cambridge University Press.

Sims, Andrea & Jeff Parker. 2016. How inflection class systems work: On the informativity of implicative structure. *Word Structure* 9.2. (Special issue on Word and Pattern Morphology edited by Farrell Ackerman and Robert Malouf), 215–239.

Šipka, Milan. 1999. Korelacije značenja i oblika duge i kratke množine imenica muškog roda u srpskom i/ili hrvatskom jeziku. In: Branko Tošović (ed.) *Die grammatischen Korrelationen* (GraLiS-1999), 165–176. Graz: Institut für Slawistik der Karl-Franzens-Universität.

Spencer, Andrew. 1991. *Morphological Theory: An Introduction to Word Structure in Generative Grammar*. Oxford: Blackwell.

2012. Identifying stems. *Word Structure* 5.1 (special issue on *Stems* edited by Olivier Bonami), 88–108.

2013. *Lexical Relatedness: A Paradigm-Based Model*. Oxford: Oxford University Press.

Stebbins, Tonya N. 2011. *Mali (Baining) Grammar*. (Pacific Linguistics, 623.) Canberra: Research School of Pacific and Asian Studies, Australian National University.

Stebbins, Tonya N. & Julius Tayul. 2012. *Mali (Baining) Dictionary: Mali-Baining Amēthamon Angētha Thēvaik*. (Asia-Pacific Linguistics Open Access Monographs, 001.) Canberra, ACT: Asia-Pacific Linguistics.

Sterner Joyce K. 1987. Sobei verb morphology reanalyzed to reflect POL studies. *Oceanic Linguistics* 26.1,2. 30–54.

Strange, David. 1972. *Dano noun inflection.* Ms, Summer Institute of Linguistics. Online: www.sil.org/pacific/png/abstract.asp?id=51968.

Stump, Gregory T. 2006. Heteroclisis and paradigm linkage. *Language* 82.2. 279–322.

2012. The formal and functional architecture of inflectional morphology. In: Angela Ralli, Geert Booij, Sergio Scalise, & Athanasios Karasimos (eds) *Morphology and the Architecture of Grammar: On-line Proceedings of the Eighth Mediterranean Morphology Meeting (MMM8), Cagliari, Italy, 14–17 September 2011*, 255–271. Online: http://lmgd.philology.upatras.gr/en/research/downloads/MMM8_Proceedings.pdf.

2015. Inflection classes. In: Matthew Baerman (ed.) *The Oxford Handbook of Inflection*, 113–139. Oxford: Oxford University Press.

2016. *Inflectional Paradigms: Content and form at the syntax-morphology interface.* Cambridge: Cambridge University Press.

Stump, Gregory T. & Raphael Finkel. 2013. *Morphological Typology: From Word to Paradigm.* Cambridge: Cambridge University Press.

Švedova, N. Ju. (ed.) 1980. *Russkaja grammatika: I: Fonetika, fonologija, udarenie, intonacija, slovoobrazovanie, morfologija.* Moscow: Nauka. [relevant section written by V. A. Plotnikova]

Swan, Oscar E. 1988. *Facultative Animacy in Polish: A Study in Grammatical Gender Formation* (Carl Beck Papers in Russian and East European Studies, no. 606.) Pittsburgh.: University of Pittsburgh, Center for Russian and European Studies.

2002. *A Grammar of Contemporary Polish.* Bloomington, IN: Slavica.

Terrill, Angela. 2003. *A Grammar of Lavukaleve.* Berlin: Mouton de Gruyter.

Timberlake, Alan. 2004. *A Reference Grammar of Russian.* Cambridge: Cambridge University Press.

Traill, Anthony. 1994. *A!Xóõ dictionary.* (Quellen zur Khoisan-Forschung/Research in Khoisan Studies, 9.) Köln: Rüdiger Köppe.

Trask, Larry. 1997. *The History of Basque.* London: Routledge.

Tucker, Archibald N. 1940. *The Eastern Sudanic Languages, Volume 1.* London: Oxford University Press.

Uredništvo [Našeg jezika]. 1949. Mrazovi: mrazevi; nosovi: nosevi i sl. *Naš jezik* (Belgrade), I/1–2, 53–54.

Velie, Daniel & Virginia Velie. 1981. *Vocabulario orejón.* Lima: Ministerio de Educación and Instituto Lingüístico de Verano.

Veríssimo, João & Harald Clahsen. 2014. Variables and similarity in linguistic generalization: Evidence from inflectional classes in Portuguese. *Journal of Memory and Language* 76. 61–79.

von Waldenfels, Ruprecht & Maciej Eder. 2016. A stylometric approach to the study of differences between standard variants of Bosnian/Croatian/Serbian, or: Is the *Hobbit* in Serbian more Hobbit or more Serbian? *Russian Linguistics* 40. 11–31.

Vuković, Jovan. 1949. *Mrazovi: mrazevi; nosovi: nosevi i sl. Naš jezik* (Belgrade) I/1–2. 42–43.

Vukušić, Stjepan, Ivan Zoričić & Marija Grasselli-Vukušić. 2007. *Naglasak u hrvatskome književnom jeziku (Velika hrvatska gramatika: IV).* Zagreb: Nakladni zavod Globus.

Wacke, K. 1930–1931. Formenlehre der Ono-Sprache (Neuguinea). *Zeitschrift für Eingeborenen-Sprachen* XXI. 161–208.

Wegera, Klaus-Peter & Hans-Joachim Solms. Morphologie des Frühneuhochdeutschen. In: Werner Besch, Anne Betten, Oskar Reichmann & Stefan Sonderegger (eds) *Sprachgeschichte: Ein Handbuch zur Geschichte der deutschen Sprache und ihrer Erforschung*, 1542–54. Berlin: Walter de Gruyter.

West, Birdie & Betty Welch. 2004. *Gramática pedagógica del tucano*. Bogotá, D.C., Colombia: Fundación para el Desarrollo de los Pueblos Marginados.

Williams, Edwin. 1994. Remarks on lexical knowledge. *Lingua* 92. 7–34.

Worth, Dean S. 1966. On the stem/ending boundary in Slavic indeclinables. *Zbornik za filologiju i lingvistiku* 9. 11–16.

Wurzel, Wolfgang U. 1984/1989. *Flexionsmorphologie und Natürlichkeit*. Berlin: Akademie-Verlag. [Translated by Manfred Schentke: Wolfgang U. Wurzel. 1989. *Inflectional Morphology and Naturalness* (Studies in Natural Language and Linguistic Theory). Dordrecht: Kluwer. [Page references to the 1989 translation.]

1990. The mechanism of inflection: lexicon representations, rules, and irregularities. In: Wolfgang U. Dressler, Hans C. Luschützky, Oskar E. Pfeiffer, & John R. Rennison (eds) *Contemporary Morphology*, 203–216. Berlin: Mouton de Gruyter.

Yu, Alan C. L. 2007. *A Natural History of Infixation*. Oxford: Oxford University Press.

Zaliznjak, Andrej A. 1967/2002. *Russkoe imennoe slovoizmenenie* Moscow: Nauka. [Reprinted in: Andrej A. Zaliznjak. 2002. *Russkoe imennoe slovoizmenenie: s priloženiem izbrannyx rabot po sovremennomu russkomu jazyku i obščemu jazykoznaniju*, 1–370. Moscow: Jazyki slavjanskoj kul´tury.] [Page references to 2002 edition.]

Zwicky, Arnold M. 1992. Some choices in the theory of morphology. In: Robert Levine (ed.) *Formal Grammar: Theory and Implementation*, 327–371. New York: Oxford University Press.

Zwicky, Arnold M. & Geoffrey K. Pullum. 1983. Phonology in syntax: The Somali optional agreement rule. *Natural Language and Linguistic Theory* 1. 385–402.

Author Index

Language Index

Subject Index

Lightning Source UK Ltd.
Milton Keynes UK
UKOW01n1022260617
304031UK00010B/137/P

9 781107 120648